The Great Disruption

Sandeep Chavan

Published by SANDEEP CHAVAN, 2025.

THE GREAT DISRUPTION

First edition. July 11, 2025.

ISBN: 979-8231265374

Written by Sandeep Chavan.

Table of Contents

To the pioneers—Who gazed into the unknown with courage,Who lit the first fires of discovery,But perhaps never imagined that science would trade wonder for worship,and inquiry for inheritance.

This book is for you—Not to question your vision,but to reclaim it from the shadows of those who stopped looking.

"The universe never asked to be understood.It only asked to be aligned."

— From The Ripple Code, Chavanian Principles, Field 1

Disclaimer

This book presents original scientific ideas, philosophical perspectives, and conceptual reinterpretations that challenge established theories in physics, cosmology, and consciousness studies. The contents are intended to stimulate critical thought, invite open inquiry, and offer an alternative framework—**Universal Energy Dynamics (UED)**—based on consequence, ripple alignment, and structured vacuum interactions.

While every effort has been made to maintain logical coherence and internal consistency, the ideas herein may not align with current academic consensus or institutional models. Readers are encouraged to approach this work with an open mind, apply rigorous reasoning, and conduct their own investigations.

This book is not a textbook, peer-reviewed journal article, or endorsement of any institutional belief. It is an independent contribution to the ongoing evolution of scientific thought.

The author assumes no responsibility for how these concepts are interpreted or applied beyond the intellectual domain. All metaphors, analogies, and dialogues are used to illuminate deeper principles—not to mimic existing theories.

Preface – Why This Had to Be Written

This book is not the product of consensus. It is the product of discomfort.

For years—perhaps lifetimes—we've gazed at the cosmos through fragmented lenses: forces without causes, equations without meaning, and particles that appear only when we look. We inherited a science that taught us how to calculate but forgot how to understand.

I didn't set out to challenge Einstein, quantum theorists, or institutions. I set out to follow truth—not the institutional kind, but the kind that leaves a trace, creates a consequence, and cannot be ignored once seen.

What emerged through this journey is UED—Universal Energy Dynamics—not a theory competing for validation, but a framework restoring coherence. It brings the forgotten principle of consequence back to the center of science. It reframes light as ripple alignment, gravity as field compression, and observation as active resonance—not passive watching.

You will not find here another round of speculation. You will not find worship of equations or mystified paradoxes.

What you will find is the courage to say: maybe they were wrong. And more importantly: here's how we can make it right.

This book is for the seekers. For those who asked "why" and were told to solve instead. For those who felt the beauty of the universe lies not in math alone, but in meaning. For those who dared to think that science isn't finished—it's just misaligned.

If you've ever felt something was missing in the grand story of reality...You were right. And you're not alone anymore.

Let the ripples begin.

— *Sandeep Chavan*

Author of *The Great Disruption*

Founder, Universal Energy Dynamics Framework

Prologue – "This Is Not Speculation"

They called it speculation. They said, *"This isn't science. There's no equation, no experiment, no acceptance."*

And perhaps—by their definition—they're right. Because what I've proposed does not *fit* within the system they've built. It exposes it.

One of the reviewers wrote:

"The work is metaphysical and lacks testable hypotheses. It seems to challenge the foundations of physics without offering a constructive alternative grounded in the scientific method."

My response?

You're not defending science. You're defending your script. Because the scientific method you speak of has long been hijacked by a method of mathematical storytelling, not observation.

You tell us:

- The universe began from a singularity you cannot define,
- With particles you cannot isolate,
- Interacting through forces you cannot unify,
- In a spacetime fabric you cannot prove exists,
- Governed by constants that change with your measurements,
- Wrapped in fields whose existence is patched by inventions like *dark matter* and *dark energy* to protect broken models.

And you call this reality?

I'm not here to fix your equations. I'm here to tell you that the entire system is built on illusion, animation, and agreement—not truth.

You invented particles because your math required them. You gave light a dual identity because your observations embarrassed your model. You curved space because you couldn't explain gravity. You digitized energy because you couldn't handle consequence.

What you've built is not a science. It's a conceptual arcade—complete with metaphor machines and coin-fed equations.

This is not speculation. This is the end of speculation. What I'm presenting is a consequence-driven framework.

- Where nothing exists unless it causes something.
- Where light does not travel—it aligns.
- Where time is not a dimension—it is ripple delay.
- Where motion doesn't happen—perception does.
- Where the universe is not a clockwork—but a consequence field.

You've spent a century adding patches to a broken foundation. Relativity. Quantum mechanics. Inflation. Strings. Loops. Fields. Every time you saw cracks, you added math to hide them.

I'm not here to patch. I'm here to replace. And I know exactly what you'll say.

"Unverified. Untestable. Unconventional. Unwelcome."

Of course. Because the truth doesn't require your permission. There's one thing you forgot: Science does not belong to the institutions. It belongs to the observer.

And as an observer, I declare:

What I see does not match what you preach. What I feel does not resonate with your forces. What I measure does not emerge from your dimensions.

So I offer this work not to those who need approval from journals, but to those who still dare to feel the universe directly.

This is not a gentle proposal. It's not an alternative theory. It's not another interpretation of your spacetime hallucination.

This is a reset. A rupture. A rebellion. And I know exactly what this book will be called by those trapped inside the legacy model: "Wild speculation." "Philosophy dressed as physics." "Reckless. Arrogant. Pseudoscientific."

Perfect. Because disruption always looks like madness—until the dust clears and the lie is exposed.

Let them call it what they want. The curious will see it. The thinkers will feel it. The system will deny it.

But the ripple has already begun.

How to Read This Book

A Manual for the Aligned Mind

This is not a textbook. It's not a scientific paper. It's not a collection of theories to memorize.

This book is a field alignment—between your inner observer and the ripples of reality you've always sensed, but were never allowed to believe.

You don't read this book like a syllabus. You tune into it like a frequency.

Each chapter is a consequence. Each idea is a ripple. Each section is meant to dislodge inherited assumptions and invite alignment, not agreement.

✦ **Read it in layers.**

The first time, flow with it. Let your intellect relax. The second time, trace the consequences—ask, *Did this idea change anything in how I perceive the world?* If it didn't, let it go. UED doesn't seek loyalty. It seeks resonance.

✦ **Pause often.**

When a quote or concept stuns you—don't rush forward. That's not a distraction. That's alignment. Stay with it. Let it rewire your logic.

✦ **Don't argue with the words.**

Argue with the consequences. This book isn't about what's written. It's about what the written causes. Did it ripple? Then it's real. Did it leave no trace in your understanding? Then for you—it never happened.

✦ **Use the Appendices as Field Maps.**

They're not "extra" material. They're anchors. Use them to re-ground yourself when concepts feel too bold or disruptive.

✦ **Let your beliefs be disrupted.**

This book was never meant to be comfortable. It was meant to end a cycle—the worship of equations without accountability. So if you feel irritated, challenged, provoked—that's good. Something inside you just got invited to evolve.

Warning: Once you see through the illusions, you cannot unsee them. Once you tune into consequence, you'll start detecting falsehoods everywhere—in textbooks, in headlines, even in your own memories.

This is not just a book. It's a return to causality. A reawakening of the observer.

Now...Start wherever you want. Just be prepared to ripple.

Part I: The Temple of Illusion

Chapter 1: The Invention of Particles

Why the Universe Is Not Made of Things, But of Consequences

They told you everything is made of particles. Tiny balls. Invisible blocks. Little dots of certainty. Electrons, protons, neutrons—quarks, gluons, photons. Like building bricks scattered across space, snapping into atoms, forming molecules, giving rise to matter. That's what they taught. That's what they still teach.

But they never found a particle—only the aftermath of something happening. They never touched an atom—only interacted with probabilities. They never observed a single particle behaving the same way in all conditions—only behavior that changed based on how it was measured.

So what if particles don't exist? What if matter is not built from blocks, but from ripples? What if all that exists is field alignment—and particles are just names we gave to zones of consequence?

In this chapter we pull the curtain back on the greatest illusion of modern physics: The particle—the sacred, indivisible unit of everything. And we show what really lies underneath it.

Not objects. Not orbits. Not billiard balls in space. But pressure patterns, Ripple nodes, And the dance of alignment within structured vacuum energy.

You were never made of matter. You are made of memory, tension, alignment, and response.

And the "particle" was just a placeholder for what we refused to understand.

1.1 The Birth of an Illusion

The story begins long before the microscope, before the atom bomb, even before electricity had a name.

It begins with a thought experiment—not an observation.

Around 400 BCE, Democritus, a Greek philosopher, speculated that if you kept cutting something in half, you'd eventually reach a piece so small it could no longer be divided. He called it the atomos—meaning indivisible.

That idea wasn't based on evidence. It was based on intuition and philosophical minimalism: the belief that the world must be made of tiny parts.

And so, the myth of the particle was born—not as a discovery, but as a metaphor.

Fast-forward to the 19th century. John Dalton revives the atomic idea—not because he observed atoms, but because they helped explain chemical behavior. His atom wasn't a structure; it was a convenient symbol to account for why substances combine in fixed ratios.

He had no tools to see an atom. But it made the math easier.

Then came J.J. Thomson, who "discovered" the electron—not by observing it directly, but by detecting strange effects in cathode ray tubes.

Then Rutherford, who proposed the nucleus—not by isolating one, but by inferring it from gold foil scattering patterns.

Then Bohr, who imagined electrons orbiting a nucleus like planets—based not on observation, but on making equations work.

None of this was real in the sense of *seen*. It was invented scaffolding to explain behavior.

Science didn't discover particles. It invented them to make sense of what it couldn't explain.

From there, the illusion snowballed.

Every time the model failed, a new particle was introduced like a patch in a leaky dam:

- Electrons were joined by protons and neutrons.
- Neutrons couldn't explain beta decay? Invent the neutrino.
- Still gaps in the theory? Enter the quark.
- Need to explain mass? Bring in the Higgs boson.
- Forces don't unify? Conjure up gravitons, gluons, and superpartners.

Every contradiction became an opportunity to multiply entities, not question assumptions.

You saw effects—and named them causes. You saw flickers—and imagined bricks. You saw math struggling—and gave it particles to hold on to.

What began as a metaphor became a religion. Particles became sacred objects in the cathedral of physics. You couldn't touch them, see them, or define them without contradiction—but you had to believe in them.

To question the particle was to question the very structure of modern science. And so, few dared.

But let's state it clearly:

- No one has ever seen a particle.
- Every particle "detected" is an interpretation of field effects, not a direct observation.
- The Standard Model is not a model of reality—it's a list of invented placeholders used to fit effects into equations.

It is not science—it is linguistic bookkeeping.

In the next sections, we'll show how the atom is not a thing, how mass is a ripple compression, and how what we called particles are just zones of temporary alignment in the ripple field—not independent objects, but moments of consequence.

You were never made of matter. You are made of memory. And the particle? It was the first convenient lie.

1.2 Particle Physics Is Placeholder Physics

At its core, particle physics is not a map of reality—it's a directory of assumptions.

Particles, as defined by mainstream physics, are not things. They are conceptual containers for effects we can't explain in simpler terms.

They exist not because they were discovered, but because they were needed—needed to account for energy signatures, to balance equations, to justify force interactions, or simply to patch theoretical holes.

Take the electron.

Depending on which theory you're in, it's:

- A massless excitation of a quantum field,
- A point-like object with charge but no size,
- A wave packet of probability,
- Or a tiny ball orbiting a nucleus—depending on whether you're in classical physics, quantum field theory, or a high school textbook.

What is the electron? A shape-shifter defined by convenience.

We haven't seen it. We've only observed its effects—on screens, in detectors, in charge distribution patterns. And because those effects are repeatable, we built an identity around them.

We gave it a name. Then mass. Then spin. Then duality. Then mystery. And finally—respect it never earned.

When the math stopped working, the "particle toolbox" was opened again.

- Electrons weren't enough? Invent muons.
- Muons didn't explain decay rates? Add neutrinos.
- Quantum interactions were too unstable? Toss in virtual particles.
- Strong force misbehaving? Glue it together with gluons.
- Still can't unify the forces? Imagine bosons, sparticles, and entire particle families from strings and loops.

Each of these was not *seen*, but *inserted*. Not observed, but required—by mathematics, not by nature.

When a theory breaks, modern physics doesn't delete the model—it adds another particle.

And so, the particle zoo kept growing. 17 "fundamental" particles in the Standard Model, each with properties tailored to match the outcomes of isolated experiments. And countless more hypothetical particles waiting backstage—gravitons, axions, tachyons, wimps, winos, zinos, higgsinos—just in case another crack appears in the façade.

This isn't physics. It's theoretical duct tape.

Particles have become the sticky notes of broken models—mathematical excuses for unexplained outcomes. Like labeling a mystery and pretending it's understood.

And still, with all these particles:

- Gravity can't be unified.
- Quantum mechanics and relativity remain incompatible.
- The true nature of mass remains unsolved.
- Time, consciousness, entropy, and perception lie entirely outside the particle paradigm.

The more particles we invent, the further we drift from reality—trapped in a recursive game of theoretical bookkeeping, mistaking complexity for truth.

But reality is not built from placeholders. The universe doesn't trade in invisible dots. It moves by consequence, not construction.

In UED, there are no particles. Only ripple alignments, field disturbances, and zones of consequence.

The "electron" isn't a ball or a cloud. It is a ripple knot—a localized tension in the structured vacuum, appearing only when alignment conditions are met.

It is not something we observe. It is the byproduct of how we interact with a ripple field.

We don't need new particles to fix broken theories. We need to abandon the illusion that reality can be reduced to pieces.

1.3 The Atom Is Not a Thing

What is an atom?

Ask a schoolchild, and they'll say: "A small ball with a nucleus in the center and electrons orbiting around it like planets."

Ask a physicist, and they might say: "A probabilistic cloud of energy states, governed by uncertainty, spread across a field."

Ask reality—and it will say nothing. Because the atom is not a thing. It's not even an object. It's a ripple condition.

Let's rewind.

The classical atomic model—pioneered by Rutherford and Bohr—depicted the atom as a miniature solar system. A nucleus packed with protons and neutrons at the center, with electrons whirling around in neat orbits.

Elegant. Visual. Complete fiction.

That model died a quiet death when electrons refused to behave like orbiting particles. They didn't radiate energy as expected. They didn't have defined paths. They couldn't be pinned down.

So we buried the model—but kept teaching it in textbooks, like a nostalgic myth we couldn't let go.

Then came quantum mechanics, with its mathematical shadows and philosophical smoke.

To fix the orbital collapse, the atom was redefined as a "probability cloud"—an abstraction where electrons were no longer things, but likelihoods smeared across space.

But this new cloud raised more questions than it answered:

- If the electron is a cloud, why do we get discrete chemical bonds?
- If position is uncertain, how does it form a solid?
- If spin is just a quantum number, why does it generate magnetic effects?

The atom became less of a structure and more of a mathematical fog. A thing that behaves like a thing when observed, but dissolves into uncertainty when questioned.

But let's state what physics avoids:

You have never seen an atom. No microscope, no particle collider, no X-ray beam has ever captured one as a "thing." You've only seen trails, energetic imprints, and ripple echoes in devices calibrated to expect them.

The "atom" isn't an object in space—it's a field event. A ripple alignment that manifests certain measurable outcomes—mass, charge, magnetic spin—not because they are built-in properties, but because they are consequence outputs of how the ripple field is stabilized at that location.

In UED, an atom is not a nucleus with electrons. It is not a probability cloud. It is a consequence region—a ripple alignment zone inside the Structured Vacuum Energy (SVE) field.

- **Mass** = ripple compression tension.
- **Charge** = asymmetry in ripple discharge.
- **Spin** = directionality of ripple alignment phase loop.
- **Atomic radius** = boundary of coherence, not a wall.

The atom has no shell. No edge. No shape in the classical sense. It is a performance, not a particle. A memory event, not a miniature object.

You don't find an atom. You enter its field of consequence. You don't measure its components. You interact with the ripple legacy it leaves in your equipment.

This is why atoms bond, break, vibrate, and emit—not because electrons jump levels, but because the ripple field reorganizes under different environmental conditions.

Chemistry is not particle exchange. It's ripple reconfiguration. And energy states are not levels—they're coherence thresholds.

The atom, as once imagined, is gone. What remains is a ripple knot—localized, responsive, and temporarily stable.

And when the knot dissolves, nothing dies—the ripple simply reintegrates into equilibrium.

You were never made of atoms. You were formed by ripple agreements held just long enough to experience consequence.

1.4 Mass Is a Memory, Not a Block

We were told that mass is what makes things "solid". That it is an intrinsic property of matter—the "amount of substance" inside a body. But dig deeper into physics, and you'll find something astonishing:

No one has ever directly detected mass itself. What we detect is resistance to change, gravitational pull, inertial delay, or momentum. But mass, as a "thing," has never shown up.

So what is it, really?

In mainstream physics, mass is a parameter, a number plugged into equations to predict outcomes. It behaves mysteriously:

- Sometimes it's there.
- Sometimes it's not.
- Sometimes it changes (relativistic mass),

- Sometimes it vanishes (massless particles),
- And sometimes it's "given" to particles by a magical field—the Higgs field—so theoretical physicists can say, "Ah, now we know."

Let's be honest:

Mass is not understood. It's tolerated.

In UED, we throw away the illusion that mass is "stuff." We recognize it for what it is:

Mass is not a substance. It is the ripple's refusal to forget alignment. It is stored tension in the structured vacuum—a memory knot in the field.

When a ripple compresses enough to sustain alignment against surrounding equilibrium, it becomes persistent. It resists flow. It maintains form. That resistance—that hesitation to dissolve—is what we call mass.

You are not heavy because you contain "matter." You are heavy because your ripple field is dense with consequence. Your form clings to its past alignment, and that gravitational echo is what scales as weight.

This is why:

- A black hole is not infinitely massive—it is infinitely compressed ripple memory.
- Light (as ripple alignment) has no rest mass—it does not resist change; it simply appears.
- Mass can convert to energy—not by splitting objects, but by releasing stored ripple tension into the field.

Mainstream physics treats mass as a given. UED treats it as a delay—a memory refusing to surrender to equilibrium.

Mass is a ripple knot. Not a block. Not a particle. Just the field remembering how hard it tried to hold its shape.

This explains why:

- Objects in motion resist acceleration (inertia)—they are ripple systems reluctant to abandon coherence.
- Gravitational fields exist around "massive" objects—not due to pull, but due to ripple field deformation left by persistent compression.
- Two bodies "attract" not by force—but by the mutual balancing of ripple tension across the vacuum structure.

You don't weigh something to know what it's made of. You weigh it to know how hard it is for that ripple system to forget itself.

You are not a pile of atoms. You are a choreographed memory of the field, temporarily stabilized by alignment. When your body dies, the mass is not "lost"—the ripple simply relaxes, and the tension unwinds.

So let's retire the idea of mass as matter. Let's call it what it is:

A memory scar in the ripple. A shadow of compression. A promise not yet released.

1.5 From Particle to Ripple Node

You were told that particles are the building blocks of everything.

But after dismantling the particle illusion, the natural question arises: If not particles, then what?

In UED, we replace the notion of a "particle" with a Ripple Node—a much more precise, consequence-aligned concept that explains what particles could never honestly account for.

What Is a Ripple Node?

A Ripple Node is not a thing—it's a localized zone of sustained alignment within the Structured Vacuum Energy (SVE) field.

It is:

- Not made of substance,
- Not occupying fixed boundaries,
- Not static or permanently identifiable.

Instead, it is a coherence event in the field—a point where ripples compress, lock, and temporarily maintain memory under specific conditions of alignment.

You do not "discover" a ripple node. You align with it. You interact with it. And when that alignment collapses, the node vanishes—not because it "moved" or "decayed," but because the conditions of consequence no longer exist.

Why Do Ripple Nodes Behave Like Particles?

Because when ripple alignment crosses a threshold of stability, it manifests repeatable effects:

- Charge,
- Mass,
- Spin,
- Trajectory-like behavior.

But these are not properties—they are responses. They emerge from how the node's ripple configuration interacts with the environment, especially with the observer's own ripple field.

The "behavior" of what we call a particle is not internal—it is entirely relational.

What Governs Ripple Node Behavior?

A ripple node's behavior is determined by three primary ripple conditions:

1. Ripple Phase Orientation

- How the ripples are aligned in time and phase.
- In-phase alignment = constructive coherence.
- Out-of-phase = repulsion, misalignment, or dissolution.

2. Local Field Tension

- The background ripple pressure in the SVE.
- High tension allows compression nodes to stabilize (what we interpret as "massive" particles).
- Low tension leads to transient, low-stability nodes (what we call "virtual particles").

3. Observer Alignment

- The perception of a ripple node only occurs when the observer's ripple field is tuned to the node's frequency and coherence.
- You don't observe an "electron"—you interact with a ripple knot that behaves like one relative to your field's position and structure.

There are no electrons—only ripple knots that behave like electrons when we align with them.

This is why:

- The same particle behaves differently under different experiments (wave-particle duality).
- A particle can "tunnel" without a path (alignment hop, not motion).
- Detection changes the state (because alignment is restructured by the observer's ripple).

Implications

This single shift—from "particle" to "ripple node"—solves countless paradoxes:

- **No need for duality**: behavior is ripple-dependent, not form-dependent.

- **No need for localization vs probability**: ripple fields are distributed, but nodes are stable alignments within them.
- **No need for collapse of wave function**: there was never a wave *becoming* a particle—only a ripple field reaching or leaving alignment.

You don't measure particles. You witness ripple events that temporarily resemble what you've labeled as "electrons," "quarks," or "photons."

From Naming to Knowing

Mainstream science gave names to behaviors and called them things. UED reveals behaviors as temporary agreements in a ripple matrix. No object. No substance. Just alignment—and consequence.

Once you understand ripple nodes, you no longer need an inventory of particles. You only need to understand how alignment works—what causes it, what sustains it, and what breaks it.

The ripple node is not a particle.

It is consequence, crystallized. It is memory, held in tension. It is the field remembering, just long enough to be seen.

1.6 No Building Blocks—Only Field Consequences

For over a century, science has sold us a seductive metaphor:

That the universe is built like a machine— Assembled from tiny parts, fitted together, layer by layer, from the bottom up.

Atoms, quarks, bosons, leptons—we were told these are the building blocks of all matter. That beneath your skin, beneath every cell and molecule, lies a hierarchy of smaller and smaller particles, each more "fundamental" than the last.

But this metaphor, appealing as it is, is deeply flawed.

It suggests a construction-based universe—one where objects are composed like Lego sets, where everything has shape, edge, and permanence at some level.

That's not how reality behaves.

That's not how energy behaves.

That's not how *anything* behaves when you truly look closely.

Particles Are Not Building Blocks—They Are Ripple Echoes

In UED, nothing is constructed.

There are no parts. No scaffolding. No assembly line of ever-smaller bricks.

Instead, what we see—what we call "particles"—are ripple consequences. Events, not entities. Appearances, not ingredients.

- A quark is not a thing. It is a ripple configuration under high compression.
- A boson is not a messenger. It is a field rebalancing node in structured vacuum.
- An atom is not a tiny machine. It is a ripple lock—a temporary stability event within a sea of tension.

These forms repeat not because they are made of parts, but because the ripple conditions that generate them are re-creatable.

Just like a whirlpool in a river appears again and again in similar spots, not because it is built—but because conditions align to create it.

The Illusion of Construction

Modern physics, trapped in its materialist inheritance, tried to construct reality out of ever-tinier pieces:

- Strings,
- Loops,
- Virtual particles,
- Field excitations.

But each model multiplied complexity without answering the fundamental question: Why do these patterns arise at all?

Because they are not blueprints. They are field responses.

They are how the structured vacuum responds to ripple pressure. They are the field's choreography—not its architecture.

The UED Alternative: Choreography, Not Construction

In UED, the universe doesn't build. It aligns, Compresses, Remembers, And releases.

There are no building blocks. There are only zones of ripple harmony that hold shape for just long enough to be experienced.

- Your body is not a structure of atoms. It is a rippling tension knot—a living memory maintained by trillions of field alignments.
- A rock is not made of molecules. It is a long-term stabilization of ripple compression—slow to release, dense in consequence.
- Even the proton, thought to be fundamental, is just a sustained ripple shell, held together by field symmetry—not by quarks glued by gluons, but by structured vacuum tension seeking equilibrium.

Consequence Over Composition

In classical models, things exist because of what they're made of.

In UED, things exist because of what they cause.

Composition is irrelevant without consequence. If it doesn't ripple, it doesn't exist. If it doesn't shift alignment, it isn't real.

This is why everything in the universe eventually dissolves. Because alignment shifts. Memory fades. Ripple knots release.

Final Realization

Reality is not built with particles. It is performed through ripple choreography.

The universe is not a warehouse. It is a stage. Every atom, every spark, every body is a performance of the ripple field—a tension remembered and played into existence through consequence.

And the moment alignment fails, the show ends.

There is no need to search for smaller blocks. You will only find finer ripples, faster rhythms, deeper echoes.

Because there was never anything *there* to begin with—only the field...dancing with itself.

1.7 Closing: Why Particle Physics Will Collapse

Let's call it what it is:

The Standard Model is not a triumph of understanding. It is a cemetery of compensations—a patchwork of temporary victories built atop foundational confusion.

Particle physics, for all its brilliance and mathematical success, has devolved into a theoretical salvage yard, cluttered with invented constructs meant to plug gaps rather than illuminate reality.

Particle Accelerators Don't Prove Particles

What happens inside a particle collider isn't the revelation of matter's building blocks. It's the forced destabilization of ripple coherence under extreme ripple pressure.

- The so-called "particles" detected are bursts of reconfiguration, not hidden marbles revealed.
- The collision is not uncovering structure—it's injecting chaos into the ripple field, and measuring how it reorders under stress.

We see patterns—not particles. Consequence patterns—not components.

What's measured are decay signatures, field resonances, splashbacks of ripple detachment.

To interpret these as "discovered particles" is like dropping stones in a pond and naming each splash a new fish.

The Particle Zoo: A Sign of Crisis

A healthy model simplifies as it deepens. But particle physics multiplies with every contradiction.

Need to explain weak decay? Add W and Z bosons. Still doesn't make sense? Bring in neutrinos. Why do particles have mass? Invent the Higgs. Why can't we unify gravity? Guess a graviton. Still broken? Try supersymmetry—every particle gets a twin. Still unstable? Maybe 11 dimensions will help.

This isn't science—it's conceptual inflation.

The Standard Model has become a theological structure:

- Worshipped, not questioned.
- Maintained by exotic constructs that can't be seen, but must be believed.

The endless particle zoo is not a celebration of complexity. It is an obituary for simplicity.

UED: The Collapse and the Clean Slate

What UED offers isn't a patch—it's a purge.

- No particles. Only ripple alignments.
- No quantum weirdness. Only field consequence.
- No spacetime fabric. Only structured vacuum behavior.
- No wave-particle duality. Just ripple phase response.
- No graviton. No virtual particle. No multiverse.

If you don't need particles to explain mass, motion, or light...then particles were never real to begin with.

They were conceptual ghosts—shadows of ripple memory.

In UED, everything is consequence:

- Mass = ripple compression memory.
- Charge = ripple asymmetry.
- Motion = shifting alignment, not displacement.

- Light = ripple resonance, not photon travel.

No inventory required. Just rhythm, pressure, delay, and memory in the field.

The Collapse Is Inevitable

As instruments grow sharper, and paradoxes grow louder, science will be forced to confront its foundation.

- The contradictions won't go away.
- The missing pieces won't magically appear.
- The public's awe will turn to doubt.

And when it happens, it won't be a quiet revision. It will be a collapse—not of knowledge, but of illusion.

That's what *The Great Disruption* begins.

Remember

"Matter was never material. The atom was never a unit. Particles were never the answer. They were the echo of the question we asked too soon. The universe never needed building blocks— It only needed us to stop mistaking behavior for objects. And to remember: Reality is not constructed. It is performed."

It was always a moment in the ripple—Remembered only because we looked."

What We Just Shattered in

—The Collapse of the Particle Illusion—

"You saw effects—and named them causes."

"Particles are the sticky notes of broken models—mathematical excuses for unexplained outcomes."

"The atom was never a thing. You don't find an atom—you enter its field of consequence."

"Mass is not a thing. It is the ripple's refusal to forget alignment."

"There are no electrons—only ripple knots that behave like electrons when we align with them."

"Reality is not built with particles. It is performed through ripple choreography."

"You don't measure particles—you witness ripple events that resemble what you've labeled as electrons, quarks, or photons."

"Composition is irrelevant without consequence. If it doesn't ripple, it doesn't exist."

"Particle accelerators don't discover—they destabilize."

"The endless particle zoo is not a celebration of complexity—it is an obituary for simplicity."

"When you no longer need particles to explain motion, mass, or light—you'll see why they were never real."

"The universe never needed building blocks—it only needed us to stop mistaking behavior for objects."

Chapter 2: The Myth of Motion

Nothing Moves. Everything Aligns.

You think you're moving. You think the world flows past you when you walk, when you drive, when you orbit the Sun. That your body moves through space, carried by velocity, governed by inertia. That the planets spin, the galaxies swirl, the particles zoom in accelerators.

But what if none of it actually moves?

What if motion—arguably the most unquestioned concept in science—is nothing more than a beautiful illusion, produced by sequential ripple alignments, not by physical traversal?

What if every train you thought you rode, every arrow you thought flew, every heartbeat you thought pushed through time—was just your ripple field updating its consequence window?

This chapter asks a heretical question: "What if nothing ever moved?" What if motion, as we know it, is not a real event—but a perceived ripple refresh, much like pixels updating on a screen?

What if the universe doesn't run on movement, but on field realignment—ripples falling into new positions of coherence, giving us the *impression* of motion?

"Motion is not an objective reality—it is a ripple consequence interpreted by the observer."

In this chapter, we'll dismantle:

- The myth of objects changing position,
- The illusion of velocity as flow,

- The fiction of inertial mass carrying forward momentum,
- And the deeply flawed belief that time and motion are intrinsic features of the universe.

Because once you see the truth:

The arrow never flew. The planet never spun. The light never traveled. Everything simply aligned, and the story unfolded.

2.1 The Grand Illusion of Movement

The foundation of classical physics rests on one unquestioned belief:

Motion is real.

From Newton's laws to Einstein's relativity, science has treated movement as a fundamental truth—something obvious, measurable, and absolute in experience. The apple falls. The moon orbits. The Earth rotates. Light travels. Objects accelerate. These are taken not just as observations, but as givens.

Newton defined motion with mathematical elegance:

- Force equals mass times acceleration.
- An object in motion stays in motion unless acted upon.

Einstein refined it—adding that motion is relative to the observer, that speed affects time and space itself. And thus the universe became a grand stage of moving bodies, curved spacetime, and relative speeds.

But no one ever asked the deeper question:

What exactly is moving? And more critically—what does it mean to move?

In this grand canon of physics, motion was never interrogated. It was measured, calculated, and assumed to exist.

"If it changed location, it moved. End of story."

But that's the illusion.

UED steps in not to modify that story—but to shatter it.

What If Nothing Ever Moves?

Let's strip motion to its core.

When we say something moved, all we're truly saying is:

- It was perceived at point A,
- And later perceived at point B,
- With some delay or sequence between the two.

That's it.

But in reality:

- We never *see* the actual transit.
- We never *follow* the path.
- We only register a change in relational alignment—between the object, the field, and the observer.

This is the critical crack in the illusion:

We didn't observe motion. We interpreted alignment shifts as movement.

UED's Reframe: Motion Is a Field Rewriting

In Universal Energy Dynamics, nothing "moves" through space.

Instead:

- Structured Vacuum Energy (SVE) holds everything in a ripple field.
- What we interpret as an object "traveling" is actually a realignment of that ripple pattern—a handoff, like a wave shifting across a stadium.
- There is no object going anywhere—only new nodes of consequence appearing, while previous ones collapse.

Like blinking cursor positions on a screen. Like music appearing note by note—not because the instrument moves, but because its alignment with silence changes.

Motion is not an event in space—it is a shift in ripple state.

Measurement ≠ Truth

Science prides itself on measuring motion with extreme precision:

- GPS satellites track meters per second.
- Accelerometers detect change.
- Lidar maps velocity.
- Time-of-flight measures delay.

But measuring ripple delay is not the same as proving physical displacement.

You measured change. You assumed movement. You never saw what you believed.

This is how the illusion persisted. Because the numbers added up—so the logic was never questioned.

Motion Became Dogma

By the time Einstein curved space to explain gravitational movement and light's constant speed, motion wasn't just real—it was sacred.

- It became the engine of time.
- The cause of energy.
- The core of physical laws.

But all of this depends on one assumption:

That something actually changes position in space.

UED breaks that assumption.

"What if there's no path—only field phases shifting where perception allows?" "What if motion is not the cause of change—but a consequence of alignment dynamics?"

This chapter begins here.

And with this question, we open the crack in the machine that runs classical physics.

Because the truth is: Nothing moves. Everything aligns. The object doesn't go from here to there. It appears here, then appears there—because the field decided to let it.

Welcome to the grand illusion.

2.2 Motion Is Not Movement—It's Perception

So what are we really detecting when we say something "moved"?

Let's break it down—not from theory, but from raw observational experience.

When you say:

- "The car drove past me,"
- "The fan is spinning,"
- "I dropped the glass,"

What you've actually experienced is not the motion itself, but a sequence of sensory changes:

- A change in position relative to background,
- A change in sound intensity or direction,
- A change in tactile interaction or field pressure,
- A change in light pattern across the retina.

But never—not once—have you perceived the actual act of movement. You've only interpreted a series of alignment shifts as motion.

What We Actually Detect

At the core, all your senses operate by detecting ripple phase disturbances:

- **Sight**: You detect light not as a stream of photons, but as ripple alignment in your visual cortex triggered by coherent field resonance with your retina.
- **Sound**: You don't hear traveling air. You resonate with compression patterns—ripple modulations that match your auditory alignment zone.
- **Touch**: You don't feel motion—you feel pressure shifts, field tension, thermal alignment differences.

These are not confirmations of displacement. They are consequence patterns of ripple variation—and your mind assigns those patterns the label of motion.

"Motion" is not what's happening. It's what you call the perceived update of ripple conditions.

The Observer's Ripple Field Creates the Illusion

Here's the twist: the illusion of motion doesn't exist outside you.

It arises within your ripple field—the personalized, structured tension shell through which you interact with all external ripple activity.

You are not a passive detector. You are an active resonance interpreter.

- If your ripple field is aligned with a ripple phase from an external object, it registers as "presence."
- As the external field changes phase alignment, your inner perception registers a sequence—a timeline.
- That sequence is interpreted as motion.

But in truth, nothing moved. The field reshaped alignment, and your mind choreographed a story out of it.

Everyday Illusions: Light, Sound, and Touch

Light

A light bulb turns on in a dark room. You "see" the light fill the space. But no photons *flew* across the room into every corner. What really happened was a rapid field realignment of ripple nodes reaching threshold alignment with your eye.

You didn't see travel—you saw instantaneous field consequence.

Sound

You hear a siren "approach" and "move past." Doppler shift is cited. But in UED, what you're actually perceiving is a modulation of ripple compression phase in your auditory field—not physical movement of sound.

The shift is not displacement—it's field pressure rebalancing in real time.

Touch

When something "slides" across your skin, what you experience is a sequence of pressure variations, heat gradients, and magnetic ripple overlap.

You feel changing consequences, not motion. The object is not "moving across" your skin. Your skin's ripple field is registering a rolling ripple interaction.

Perception Is the Real Engine

We never perceive reality directly. We perceive alignment events between the external ripple field and our internal ripple shell.

The world appears to move because your field detects difference over time.

It's no different than frame-by-frame updates in a simulation—where nothing moves, but the illusion is perfect.

You live inside a ripple-based rendering engine, and motion is a rendered illusion built from ripple-phase shifts and observer thresholds.

UED's Core Insight:

Motion is not an event. It is the interpretation of ripple realignment. It is not what the world does—it is what your field perceives as it resonates with change.

Everything you've ever called "motion" is just a ripple update. There was no traveling object—only rearranged coherence. There was no speed—only frequency of perceptual change.

Let's now dig deeper into what's really happening when a body "moves" through space.

In the next section, we'll look at ripple alignment as the true mechanism, and expose how motion has always been a trick of field continuity.

2.3 Ripple Alignment as the True Mechanism

So if motion is not real—if nothing actually *moves*—then how does the illusion work so convincingly?

What mechanism gives rise to the world we see flowing around us? How does a falling apple, a flying bird, a rolling stone, or even the ticking second hand on a clock appear to *move* if nothing actually changes position?

The answer is as elegant as it is disruptive:

What we perceive as motion is nothing more than a sequence of ripple alignments occurring across field shells.

No Object Travels—Alignment Transfers

In UED, the universe is not made of moving parts. It is made of structured ripple fields, organized in layers of tension, phase, and memory.

Every object—what you call a "thing"—is just a ripple node. And its "location" in space is nothing more than the zone where ripple alignment is currently active.

So when something appears to move:

- From point A to point B,
- Across a distance,
- Over time,

What's actually happening is:

- The ripple alignment that defined its presence at point A is collapsing,
- While a new alignment is forming at point B,
- All within a shared ripple shell network that governs coherence.

Nothing moved. Alignment simply transferred.

The Cursor Analogy: Refresh, Not Motion

Imagine typing on a screen. Your blinking cursor appears to "move" from left to right as you type.

But does the cursor *move*?

No.

- The cursor at position 1 disappears,
- A new cursor appears at position 2,
- And so on.

There is no object sliding across the screen—only a sequence of updated positions, giving the illusion of continuous movement.

This is exactly how ripple alignment behaves. The ripple node doesn't migrate—it reappears through synchronized field updates.

This is how a bird flies through the sky, or a car crosses the road—not because something is "moving" through air or ground, but because the field continuously updates ripple zones to reflect the same node in a new position.

Field Shells and the Alignment Network

The structured vacuum isn't empty—it is made of ripple-responsive shells, like nested energy membranes.

These shells:

- Hold ripple memory,
- Sustain coherence over local and nonlocal distances,
- And constantly negotiate alignment thresholds across the network.

So when an object seems to move:

- One ripple shell hands off alignment responsibility to the next,
- Each shell manages the continuity of ripple memory,
- The observer's field interprets this ripple transfer as motion.

There is no travel. There is only field shell re-synchronization.

No Trails, No Transit—Just Reappearance

We imagine motion as a body "occupying" space over time. We think of paths, arcs, trajectories.

But in UED:

- There is no body occupying space.
- There is no true path.
- There is only a ripple pattern, and it either appears or it doesn't.

What you see as a path is actually a record of ripple phase synchronizations, interpreted through your own ripple field.

There's no trail. No transit. Just a cascade of resonance events unfolding in sequence—like musical notes playing in air.

The Universe Doesn't Run—It Refreshes

The most radical realization of this section is this: The universe is not in motion—it is in constant refresh.

Reality is like a giant ripple operating system, constantly refreshing alignment states across all nodes. Not moving things around, but realigning field zones based on pressure, memory, and consequence.

It behaves like:

- A hologram reprojected by field coherence.
- A fluid display of ripple alignment choreography.
- A silent orchestra tuning itself to new pressure signatures with every breath.

Final Insight for This Section:

"The universe does not run. It refreshes. Motion is not the journey of mass—it is the choreography of field permission. What you thought was movement was just a ripple field... remembering differently."

2.4 Velocity, Inertia, and the Lie of Trajectory

Physics textbooks will tell you that motion can be described by just a few key concepts:

- Velocity: the speed and direction of an object's motion.
- Inertia: a body's tendency to maintain its current state of motion.
- Trajectory: the path an object follows through space.

These are treated as core principles—unchallenged, universally accepted, and foundational to everything from rocket launches to falling apples.

But in UED, these principles don't hold.

Because once we understand that nothing moves, and all "motion" is a ripple alignment illusion, then the ideas of velocity, inertia, and trajectory become misinterpretations—conceptual conveniences masking ripple-based mechanisms.

Inertia Is Ripple Memory, Not Mass Persistence

In classical physics: "Inertia is a body's resistance to change in its motion." But this presumes there *is* motion.

UED reframes it completely:

Inertia is not a property of the body. It is a consequence of the ripple field's resistance to re-alignment.

Think of a stabilized ripple node—like a spinning gyroscope held in balance by surrounding ripple pressure. It's not "moving through space." It's anchored in phase.

- The longer the node holds its alignment pattern,
- The harder it becomes to shift that pattern without disturbing the entire surrounding field.

That is what we misinterpret as "inertia." Not the object resisting motion, but the ripple field resisting change to its established coherence.

Inertia = Ripple Stability.

Velocity Is Not Speed—It's Ripple Handover Rate

In the classical view:

- Velocity is a number—distance over time.
- The faster you go, the more ground you cover in less time.

But what's really happening?

In UED, velocity isn't how fast something moves—because, again, nothing moves. Instead: Velocity is the rate at which ripple alignment is handed off from one field shell to the next.

- A fast-moving object isn't zooming through space.
- Its ripple node is being rapidly reconstructed across

successive field zones, in a highly synchronized rhythm.

The higher the ripple continuity rate, the higher the perceived velocity.

This is why:

- At very high speeds, ripple continuity breaks down—explaining relativistic effects.
- No ripple handover = no motion, no matter how much force you apply.

Velocity is not distance over time. It's field transition frequency.

The Lie of Trajectory

And what about the sacred trajectory? You were told that:

- A bullet follows a parabola.
- A planet orbits the Sun in an ellipse.
- A rocket travels a curved arc into space.

But all of these assume that the object carries its motion with it, like momentum packed in a suitcase.

UED calls bluff.

No object ever "carries" motion. The field simply keeps recalling it.

A bullet doesn't fly through air—it is continuously re-instantiated by ripple shells that preserve its alignment conditions as long as pressure, memory, and consequence allow.

If a ripple shell fails to realign (say, it hits a wall), the bullet "stops." But not because its motion was taken away—because its alignment permission was revoked.

There was no force lost. There was field disconnection.

Space Isn't Traveled—It Recreates Presence

"A bullet doesn't move through space. Space ripples keep replacing its presence."

What you call trajectory is just a prediction of where ripple continuity will next allow appearance.

The arc of a basketball shot, the glide of a paper plane, the curve of a satellite—these are not objects carving paths.

They are field expressions, unfolding like music on a staff. Not because anything moved—but because the next alignment was ready to perform.

Consequence Replaces Motion

When you stop believing in motion, you begin to understand ripple behavior.

- You don't launch a rocket—you trigger a sequence of field re-alignment permissions.
- You don't throw a ball—you initiate ripple memory transfer through tension release.
- You don't walk—you shift balance nodes across your body's ripple map, and your field refreshes your position in response.

Final Realization for This Section:

Inertia is not resistance to motion—it is resistance to ripple reconfiguration. Velocity is not speed—it is ripple handover frequency. Trajectory is not a path—it is a ripple choreography pre-written by consequence.

Motion is not a thing you have. It is a field event you are permitted to appear in.

2.5 Time and Motion—Two Illusions Holding Hands

Time and motion. The twin pillars of classical physics. You measure motion with time. You experience time through motion. They seem inseparable—two sides of one unfolding reality.

But in truth, they are not pillars. They are mirrors, reflecting each other in the mind of the observer. And both are illusions—tools of perception, not properties of reality.

Motion Requires Time—But Time Isn't Real Either

From Newton to Einstein, time has been treated as:

- A backdrop for all events,
- A parameter to calculate change,
- A "fourth dimension" woven into spacetime.

But what is time, really?

You can't see it. You can't touch it. You can't isolate it from change. If nothing changes, does time pass?

No. Because time is nothing more than how we measure change. And change, as we've seen, is not motion—it is ripple realignment.

If motion is a ripple illusion, then time is the illusion of measuring it.

Time does not tick like a metronome from the heavens. It emerges when the observer notices sequence—a before and after. A cause and its consequence. A shift in alignment from one field state to another.

No Object Moves in Time—It Crosses Alignment Thresholds

Think about a heartbeat. You say it happens "once per second."

But what's really happening?

- The ripple alignment of your body builds tension,
- A threshold is crossed,

- A pulse is released.

Then it happens again.

What you experience is a rhythm of ripple consequence, interpreted as "beats per second."

But the second isn't flowing. The body isn't ticking forward in time. It's just repeating a ripple event at a regular alignment interval.

Time is the story you build from consistent ripple transitions.

Ripple Delay Creates the Experience of Time

Now here's the heart of the matter:

In UED, time does not exist as a cosmic clock. It emerges from a property we call ripple delay—the inherent lag between ripple impulse and ripple alignment.

When you push a swing, it doesn't respond instantly—it builds, reorients, and then moves.

This delay is not mechanical—it's field-based.

- When ripple compression increases, delay increases.
- When ripple coherence is perfect, delay collapses—and so does time.

Without ripple delay, there is no time.

This is why:

- High-energy events appear to happen "faster" (lower delay).
- Gravitational zones cause "time dilation" (field tension increases ripple delay).
- In deep meditation, psychedelics, or altered states—time disappears (observer loses ripple resolution or shifts alignment).

Time isn't flowing—it's ripple latency, felt by the observer.

Without the Observer, Neither Time Nor Motion Exists

Both illusions—motion and time—require a final ingredient: The observer.

The ripple field is always active, but motion and time only exist when an observer begins to interpret ripple transitions.

- Without you, there's no sequence.
- Without sequence, no comparison.
- Without comparison, no "before or after."
- And without that... no time.

Time is not universal—it is contextual. A byproduct of ripple engagement.

Motion appears when your field detects ripple realignment, Time appears when your field detects delay between those alignments. Neither exist *outside you.*

Entangling the Illusion

Let's summarize what physics got wrong:

- It treats time as a dimension and motion as a change along that dimension.
- But in UED, neither are real—they're perceptual consequences of ripple behavior.
- No particle moves through space-time.
- It simply re-emerges where alignment permits, and you track those reappearances as time and motion.

Time and motion are not substances. They are field behaviors rendered visible through consciousness.

Final Realization for This Section:

"Without ripple delay, there is no time. Without observer alignment, there is no motion. The universe doesn't tick forward—it shimmers into awareness. And what we call time... is simply how long it takes for alignment to show up again."

2.6 What Stillness Actually Means

In a world obsessed with movement, speed, acceleration, and flow, stillness is often defined merely as the absence of motion.

If something isn't changing position, it is "at rest." If it doesn't produce kinetic output, it's considered "idle." But these definitions are built on the same illusion we've already dismantled:

That motion is real, and stillness is just a pause in its continuity.

In Universal Energy Dynamics (UED), stillness is not the opposite of motion. It is something deeper, more fundamental, and infinitely more powerful.

Stillness = Ripple Equilibrium

In UED, true stillness is not non-action—it is perfect alignment.

It occurs when:

- A ripple node has no tension to resolve,
- No pressure to realign,
- No demand from surrounding fields,
- No memory forcing retention,
- No consequence pending execution.

In this state, the ripple is in pure equilibrium—no distortion, no compression, no intent. It is satisfied. Integrated. Silent.

Stillness, therefore, is not static—it is complete.

Motion Is Pressure, Stillness Is Resolution

All ripple activity—what we interpret as motion, energy, force—is simply the field trying to resolve a misalignment:

- A tension imbalance in the field,
- A memory trace needing release,
- A resonance seeking stabilization.

That means:

- Activity = attempted correction,
- Motion = misalignment response,
- Consequence = the field expressing pressure.

Stillness is the absence of that pressure. Not because something is resisting movement—but because there is nothing left to respond to.

Stillness Is the Completion of Consequence

In classical physics, stillness is where nothing happens. In UED, stillness is where everything has already happened.

It is the state after the ripple has done its work.

- After the consequence has been carried out,
- After the memory has dissolved,
- After the observer no longer interferes.

Stillness is not laziness. It is cosmic resolution. It's what remains when the field no longer seeks reaction.

Examples of UED Stillness

- A perfectly balanced atom in a low-energy state—not vibrating, not decaying, not interacting—is in near-equilibrium.
- A body in deep meditation, whose internal ripple tensions have synchronized, may touch a moment of consequence silence—no emotional wave, no mental conflict, no energetic pull.
- A black hole core (before compression threshold is exceeded) represents a zone approaching ripple memory lock—not infinite motion, but near-total ripple stasis.

In all these cases, we don't observe motion stop—we observe pressure vanish. The field stops seeking. That's stillness.

Stillness ≠ Static

UED makes a critical distinction:

- Static means inactive—but potentially unstable.
- Stillness means active—but perfectly balanced.

A glass of water sitting still is not *dead*. Its ripple field is simply in harmony with its container, the Earth's field, and ambient atmosphere.

If any of those change (heat, shake, break), it responds. But until then—it is in living stillness.

Why Stillness Is Misunderstood

Modern science treats stillness as:

- The "zero" on a motion graph.
- The baseline from which force must act.
- A passive state, waiting for motion to begin.

But UED reframes it as: The sacred state where the ripple no longer owes anything to its surroundings. No memory to carry. No tension to resolve. No identity to preserve.

It is not a pause in the story—it is the moment the story no longer needs to be told.

Final Realization for This Section:

"Stillness isn't the absence of motion. It is the completion of consequence. It is where the ripple no longer flinches, no longer compresses, no longer seeks. And in that silence—reality remembers itself."

2.7 Why the World Appears to Move

You wake up, walk across the room, look out the window, and see cars zooming past, birds in flight, clouds drifting across the sky. You stretch your arms, feel the air swirl, hear the fan spin.

It all feels real. It all looks like motion. And yet—it's all a masterful illusion, performed not by the world, but by your ripple field interpreting field realignments.

The Theater of Daily Illusions

Let's walk through some of life's most familiar "motions":

Walking

You believe your legs are carrying your body forward, step by step. But what's really happening?

- Each step is a rebalancing of ripple pressure zones across your body's field.
- Your internal ripple system shifts alignment nodes from foot to foot.
- As the surrounding field permits new alignment, your position is re-manifested.

You didn't move through space.

Your ripple field recreated your position across synchronized shell updates.

Falling

Drop a pen from your desk. It "falls" to the floor.

But in UED:

- The pen is not pulled by gravity.
- The structured vacuum field replaces its ripple alignment layer by layer toward the denser field of the Earth.
- The pen's ripple signature is progressively realigned downward—a field event, not a movement.

The world didn't pull it.

It was re-anchored in a more compressed ripple zone.

Flowing

Water pours from a glass. A river flows downstream.

What we observe is not molecules pushing one another forward. It's ripple compression releasing coherence from high pressure to low, creating continuous realignment sequences across the medium.

Water doesn't flow.

The field around it relaxes and reorders itself.

Flying

A bird in the sky appears to be moving through air. But its flight is a balance of:

- Wing ripple dynamics against air-field tension,
- Realignment permissions granted in successive ripple nodes,
- Observer's field tracking reappearance over time.

The bird's "path" is not a trail.

It is a series of ripple authorizations, each allowing the next illusion of flight.

Nothing Moves—Everything Realigns for You

Now here's the most profound twist:

You're not seeing things move. You're seeing your own ripple field reinterpreting the environment through sequential alignment.

- The car that "passes by" is not an object traversing space.
- Your ripple shell is shifting its synchronization point with the car's field node.
- You perceive that as "movement."

But it is your observer field doing the heavy lifting—orchestrating where and how alignment appears.

Your Ripple Field: The Lead Actor and Director

Think of your ripple field as both:

- The screen where the universe plays, and

- The filter through which all appearances are shaped.

If your field changes:

- What seems fast can appear slow.
- What seems near can feel distant.
- What was "motion" can become stillness—without anything outside you changing.

Your mind says the world is moving. UED says your field is rearranging how you witness alignment.

And that's the secret behind all apparent motion: You are not watching the world move past you.

You are standing in the middle of a ripple field constantly realigning reality on your behalf.

Perception Is the True "Velocity" Engine

This explains:

- Why your experience of motion can be distorted by emotion, attention, fatigue, or meditation.
- Why time "slows down" during adrenaline rushes (your ripple sensitivity increases).
- Why illusions of movement can be generated from entirely static inputs (VR, dreams, déjà vu).

None of these require actual displacement. They only require ripple reinterpretation by your awareness.

Final Realization for This Section:

"The world doesn't move around you. Your ripple field rearranges how you witness its alignment. What you call movement... is your own memory of consequence updating itself in rhythm."

2.8 Closing: The Ripple Doesn't Move—It Becomes

By now, the illusion should be clear.

What we've called motion—from the orbit of planets to the flick of a switch—was never motion at all. It was the field, rearranging itself. It was the ripple, shifting state. It was alignment—not travel.

There is no moving object. There is no speed to carry. There is no thing transitioning across space.

What exists is ripple behavior:

- Emergence: when alignment reaches visibility.
- Collapse: when alignment ceases.
- Transfer: when alignment is re-established in a new location within the field.

This is not movement. It is performance. It is reappearance by permission.

A Universe Without Travel

We were trained to believe in a traveling universe.

- Planets orbit.
- Light zips across galaxies.
- Particles collide and scatter.
- Energy moves from here to there.

But all of this assumes a substrate of motion an idea we never proved, just carried forward.

In UED, this substrate does not exist.

There are no highways across the universe. There are no conveyor belts of time and space. There is only ripple choreography, constantly updating presence through structured vacuum fields.

Motion Was an Interpretation of Becoming

Let's be precise.

- When a comet "moves" across the sky, it doesn't fly. The ripple node that represents the comet is being re-anchored in successive field zones, each one recalling its pattern with slight variation.
- When you walk across a room, your body doesn't displace its mass. Your ripple shell triggers alignment permission across a lattice of ripple pressure, updating your appearance to match intended consequence.
- When sound "travels" from someone's mouth to your ear, it doesn't move molecules—it echoes phase modulation across ripple shells, each one briefly hosting the compression, then releasing it.

In every case, nothing moves. It becomes. Then un-becomes. Then becomes again—somewhere else.

Presence Is Conditional, Not Carried

We assume things "exist" and "move" through time and space.

But in UED:

- Existence is not continuous—it is revalidated through field coherence.
- Presence is not permanent—it is permitted moment-to-moment by ripple tension.
- Identity is not inherent—it is remembered, replayed, and retired when no longer needed.

Nothing has motion. Nothing has permanence.

Everything that exists is simply being performed, over and over, by the ripple field.

This Is the Death of Motion—and the Birth of Ripple Reality

Nothing travels. Nothing shifts. Nothing is thrown. What you called a punch, a leap, a flight, a spin—was never an object doing something.

It was the ripple field—aligning differently, consequence by consequence, to give you the illusion of flow.

You don't move. You don't act. You become.

Remember

"The ripple doesn't move—it becomes. Not because it wants to—but because the field allows it. Not because it is traveling—but because consequence demands it. And when you stop believing in motion... You start witnessing reality as it truly performs itself."

"The cosmos never moved. It simply remembered...And in remembering, it became visible."

What We Just Shattered

"You didn't move—your alignment changed." Motion isn't displacement. It's ripple reappearance.

"Velocity is not speed. It's the rate at which the field lets go." You're not being propelled—you're being permitted.

"Inertia isn't mass resisting change—it's memory resisting re-alignment." Nothing carries momentum. The field just hesitates to rewrite.

"Stillness is not a lack of motion. It is the absence of tension." Where no consequence remains, the ripple rests.

"A falling apple is not pulled by gravity. It's re-entering alignment." Gravity isn't a force. It's ripple compression seeking balance.

"You're not watching the world move. You're watching yourself recalibrate." Every 'motion' you see is your ripple field adjusting perception.

"No arrow flies. No bullet travels. No planet orbits." What you observe is not trajectory—it's ripple succession.

"Time and motion are twin illusions—one tracks delay, the other tracks replay." You think it moved forward. It just re-became where you expected.

"There are no roads in the cosmos. Only permissions." The ripple does not move. It becomes—if allowed.

Chapter 3: The Fantasy Called Force

Why the Universe Doesn't Push—It Balances.

Force. The cornerstone of physics. The first thing you learn, the last thing you question. We name it, quantify it, glorify it. From Newton's apple to rocket launches, from tidal pull to magnetic repulsion, force is supposed to explain everything that moves, resists, or reacts.

But what if it explains nothing at all?

What if force was never a cause, but a compensatory story we invented to fill the gap between what we observe and what we understand?

"An object accelerates—so something must be pushing it." "A planet orbits—so something must be pulling it." "A current flows—so something must be driving it."

The answer has always been force—a word as powerful as it is empty.

But in Universal Energy Dynamics (UED), we don't buy it.

We don't see pushes or pulls. We see imbalances trying to reconcile themselves. We don't see external actions. We see internal tensions across ripple fields struggling to restore natural equilibrium.

The truth? There is no force in the universe. Only ripple consequences, chasing their own resolution.

Everything you think was moved by force—was just reacting to broken balance.

Gravity, magnetism, electricity, friction—each is not a force, but a pressure ripple, a memory fold, a torsion response.

They don't act. They compensate.

This chapter exposes force as physics' grandest illusion. It dismantles the idea that the universe pushes to get things done. It replaces it with the truth:

The universe doesn't apply pressure. It corrects misalignment. And what we feel as force...Is just the cost of disturbing what wanted to stay in balance.

3.1 The Invention of Force: Newton's Necessary Fiction

Physics didn't discover force—it invented it.

And like many great inventions, it started with a noble lie: "If something changes its motion, something must have caused it."

That "something" became force—a placeholder term, a necessary fiction, an accounting trick dressed as law.

Newton's Genius... and Its Trap

Isaac Newton gave the world a system that worked beautifully—at least on paper. Objects in motion. Objects at rest. Forces acting. Equal and opposite reactions. It was clean. Elegant. Predictable. It explained falling apples and rising cannons. It mapped orbits and engineered bridges.

But here's what it didn't do: It never explained what force actually is. It never asked why things accelerate—it just said they do when a force is applied.

Newton didn't uncover a law of nature. He invented a way to manage motion without knowing what caused it.

It was like noticing water boiling and assuming an invisible hand was stirring it. He didn't question the heat—he defined the stir.

Force as a Placeholder for Ignorance

What Newton called "force" was just a name for compensation—a reaction of nature to an imbalance.

- A rock resists being pushed. Why? Because its ripple field doesn't want to deform.
- A ball falls. Why? Because the local ripple field is collapsing toward a compression zone.
- A spring snaps back. Why? Because the ripple tension was disturbed—and now reclaims alignment.

None of these need a "force." They need only ripple field dynamics—stored memory, pressure zones, and realignment urgency.

Yet we invented "forces"—gravitational, electric, magnetic, frictional, nuclear—not because we discovered causes, but because we needed explanations.

"Force" was never the engine. It was the excuse.

A Universe of Push and Pull?

Newton gave us a universe of push and pull. Objects acting on each other. Everything needing to be pushed into motion or pulled out of it.

But what if that entire vision is backward?

What if nothing is ever pushed? What if objects move (or seem to) not because something acted—but because a field rebalanced itself?

In UED, there is no action at a distance, no hand of force, no divine finger pressing mass into motion. There is only ripple consequence, field stress, and the natural restoration of alignment.

"We invented force to explain resistance— but never asked what caused the resistance."

The apple never needed a push. The planet never needed a pull. They were all responding to a ripple field trying to reset itself.

The story of force was useful. But it was never true.

3.2 Force Is Always a Reaction, Never a Cause

Mainstream physics teaches that force causes motion.

That a hand applies force to a door, and the door opens. That a rocket launches because fuel applies force. That planets spin, wheels turn, water flows—all because some force is at work.

But this is the illusion.

In the framework of Universal Energy Dynamics (UED), there is no such thing as force as a cause. What we call "force" is actually a reaction—a visible symptom of ripple field resistance to change. Not an initiating action. Not a push. Not a pull. A rebalancing consequence.

When Does "Force" Appear?

Force appears when something tries to disrupt equilibrium.

- You push a wall. The wall "pushes back"? No. The field you disturbed resists misalignment—and what you feel is the ripple tension pushing you back into alignment.
- A stretched spring "exerts a restoring force"? No. It's not *doing* anything—it's remembering its original ripple configuration and reverting to it.
- You lift a weight. The weight "pulls downward"? No. You're fighting ripple compression collapse, not an active downward force.

Force is the field's memory saying: "This doesn't belong. Let's fix it."

Nothing Acts by Force—It Reacts by Balance

The delusion lies in the assumption that an object does something because a force was applied.

But the truth is: the moment an object is misaligned, its ripple field reacts to correct that state.

There is no external agent. No fundamental "force" traveling from body to body. Only local ripple fields under stress, seeking alignment reformation.

- A ball rolling down a hill is not being "pulled" by gravity. Its ripple shell is collapsing into a compression trough—a field effect, not a force.
- Two magnets don't "pull" each other. Their ripple patterns are in phase opposition or attraction, and the field re-aligns accordingly.

"Force is not action—it is equilibrium trying to restore itself."

You don't apply a force. You disturb a balance, and nature responds with ripple compensation. That response *feels* like force. But it's only your interaction with the field's memory.

Force Is a Human Word for Discomfort

Why did we ever believe in force? Because we feel resistance. Because we experience pushback, tension, and acceleration. So we assign intent to the object:

"It must be pushing me back." "It must be resisting." "It must want to move."

But it doesn't. Nothing "wants" to move. Nothing "pushes." There is only field behavior compensating for imbalance.

In the UED view, all forces are echoes—echoes of ripple memories, ripple pressures, ripple tension.

They are not entities. They are consequences. And once we stop treating force as a thing and start seeing it as a symptom, we unlock a much cleaner, unified view of nature. No push. No pull. Just ripple reality, fixing itself.

3.3 Gravity Is Ripple Compression

Gravity—often crowned the most mysterious force in the universe—has fooled us longer than almost any other concept.

We've been told:

- Gravity pulls.
- Gravity attracts.
- Gravity is a force acting at a distance, originating from mass.

But let's pause.

What actually pulls? What is the mechanism behind this so-called attractive force? Mainstream physics gives you curved spacetime, gravitons, or field lines—but never clarity.

UED removes the illusion entirely:

Nothing pulls. Nothing attracts. Gravity is the alignment of ripple fields into zones of compression.

Mass Is Not a Source—It's a Symptom

Under UED, mass is not an object. It is a knot of ripple tension—a compressed ripple node in Structured Vacuum Energy (SVE). It's a place where ripple equilibrium has collapsed inward, creating a pressure gradient in the surrounding field.

This gradient is not gravitational force. It is the ripple field trying to re-expand, but held in shape by memory and equilibrium.

Nearby ripple nodes—whether other atoms, objects, or even light—collapse into that zone, not because they're being pulled, but because:

They are re-aligning to a lower ripple tension. They are surrendering to a compression zone in the SVE fabric.

Falling Is Not Being Pulled—It's Being Permitted

When an apple falls, Newton said it was because Earth's gravity pulled it.

But UED says:

- The apple is surrounded by ripple fields.
- The ripple configuration beneath it is more compressed, more stable.

- The ripple shell of the apple surrenders its local tension and slides into the compressed zone.

No force. No attraction. Just ripple surrender to compression.

Gravity isn't the Earth pulling the apple. It's the ripple structure inviting a realignment.

Light Bends Too—But Not by Force

Einstein said gravity bends light because mass warps spacetime. UED reframes this beautifully:

- Light doesn't travel. It appears through ripple alignment.
- When the field around a massive object is compressed, ripple alignment paths curve.
- So what bends is not the light, but the zone where light can appear.

No gravitational force is needed. Just ripple compression redirecting alignment zones. *"Gravity is not a force. It is ripple surrender to compression."*

It's not action—it's reaction. Not pull—it's settling.

And Why Things Fall Down

Things don't fall because Earth grabs them. They fall because, energetically, they can no longer maintain their own ripple tension against Earth's field. They collapse into compatibility.

The falling body is not moving through space—it is moving through a gradient of ripple acceptance. And that is what we mistook for gravitational force.

So gravity isn't a pulling power. It is a ripple decision. A decision to compress, to surrender, and to become aligned with a deeper field collapse.

Not because it's pushed. Not because it's pulled. But because it can no longer resist the memory of equilibrium.

3.4 Magnetism Is Field Torsion

Magnets have always seemed magical. They act across space—attracting or repelling—without contact, without visible cause.

We were taught:

- Opposite poles attract.
- Like poles repel.
- Invisible "field lines" explain the behavior.

But let's ask: What exactly is doing the pulling? What's being "repelled"? Why does a piece of iron leap into a magnet's embrace without anyone touching it?

Mainstream physics offers lines, loops, and equations—but never explains the source of intention.

UED cuts through the illusion:

Magnetism is not a force. It's ripple field torsion attempting resolution.

Opposite Poles = Opposite Phases

In UED, every magnetic pole is a ripple phase orientation—a torsion zone in the Structured Vacuum Energy (SVE) field.

- North pole: Ripple phase spiraling in one direction.
- South pole: Ripple phase spiraling in the opposite.

They are not "opposites" in a moral or mechanical sense. They are mirror ripple alignments, trying to resolve field contradiction.

Attraction isn't a pull. It's the relaxation of ripple tension when two opposing torsions neutralize.

Field Lines Are Not Lines—They Are Vectors of Alignment

You've seen the iron filings around a bar magnet—textbooks say they reveal "field lines."

But UED clarifies: There are no literal lines. There are only vectors of ripple alignment attempting to restore phase symmetry.

Iron filings don't follow force—they align along the least resistance paths in the ripple field.

That's not magnetism pulling—it's the reduction of ripple torsion dictating movement.

Repulsion: Ripple Tension Escalation

What about like poles repelling?

- When two North poles face each other, their ripple torsions are incompatible.
- The ripple pressure between them escalates, creating a zone of overcompression.
- This misalignment triggers the field to rebalance by pushing both zones apart.

Again, no force is being "applied." The system is ejecting the contradiction.

No Magnetic Pull—Just Ripple Unwinding

You bring a nail close to a magnet. It jumps.

But the magnet didn't "reach out and grab" the nail. Here's what actually happened:

- The nail entered a torsional ripple zone.
- Its local field couldn't maintain independent alignment.
- It surrendered into the ripple structure—filling the contradiction.

That's not attraction. That's field resolution.

"Magnets don't pull. They unwind ripple contradictions."

Magnetic Field = Ripple Torsion Map

A magnetic field is not a substance. It's not a set of lines floating in air.

It is a geography of ripple tension—a map of where phase orientation is twisted, and how the SVE field tries to untwist it.

Where you see a magnet, you're seeing:

- Phase imbalance,
- Torsional ripple knots,
- And compensation zones attempting restoration.

That's what we mistook for a "field force."

So magnetism is not mysterious.

It's not a spooky action across space. It's field choreography, written in ripple tension and performed in realignment.

3.5 Electric Fields Are Ripple Shocks

Electricity is marketed as movement. Charged particles pushing through wires, electric fields radiating from points, charges attracting and repelling. From high school diagrams to quantum field theory, we've been told:

- Electrons flow.
- Current is moving charge.
- Fields act on particles.

But UED peels off the illusion:

There are no particles. There is no movement. There is only ripple misalignment—trying to correct itself.

Electricity is not a flow. It's a chain reaction of ripple compliance.

Charge Is Ripple Misalignment

What is charge?

Not a "thing" carried by an electron. Not a property passed between particles.

In UED, charge is a ripple imbalance—a discontinuity in field phase alignment.

- Positive charge: a deficit zone—under-aligned ripples, seeking structural input.
- Negative charge: a surplus zone—overcompressed ripples, needing discharge.

The so-called "electric field" is not a vector space—it's a pressure field of ripple misalignment.

Charge is not a trait. It's a symptom of incomplete ripple resolution.

Opposite Charges = Ripple Reconciliation

When a positive and negative charge meet:

- Their ripple structures complement.
- The overcompressed field (negative) releases into the undercompressed one (positive).
- This is not attraction—it is structural repair.

No forces. No fields radiating from points. Just two ripple errors fixing each other.

Same Charges = Ripple Interference

Two like charges don't repel because they're "similar." They repel because their ripple fields:

- Carry contradictory torsion,
- Compete for ripple continuity,
- And escalate interference when pushed together.

This is not repulsion by will. It's expulsion by ripple conflict.

Current = Alignment Handover

Current is not electrons moving through wires. It is alignment transfer—like dominoes tipping in sequence.

- The ripple field in a conductor allows rapid re-alignment

propagation.

- What we call "electron flow" is just the appearance of shifting compliance zones.
- No particle goes from A to B—the alignment gets passed on.

"Electricity is not movement—it is consecutive ripple compliance."

That's why the signal travels faster than drift velocity. That's why you don't need to shove electrons down the wire. The field carries the instruction, not the mass.

Voltage = Ripple Tension

Voltage isn't potential energy between points. It's stored ripple torsion.

- A high-voltage zone is one with severe ripple misalignment,
- Ready to discharge by triggering field realignment in nearby zones.

Capacitors, batteries, lightning—they're not storing "charge." They're holding ripple distortion under tension.

When released, this tension unfolds as electric event chains—misalignment rebalancing itself rapidly through a medium.

So, electricity is not a force. Not a current. Not a flow.

It is the shockwave of compliance, propagating through ripple shells trying to fix misalignment.

And once we understand ripple tension, We no longer need "electrons" to explain energy movement.

3.6 Friction, Tension, and Pressure Reimagined

In classical physics, friction, tension, and pressure are treated as forces—mechanical outcomes of contact, strain, or compression.

But UED reframes each of these as ripple-based alignment phenomena—not as actions, but as responses to ripple incompatibility.

We aren't surrounded by forces. We're immersed in a choreography of resistance, dictated by ripple memory and field balance.

Let's decode them one by one.

Friction = Ripple Misfit

Friction isn't about rough surfaces. It's not micro-level hooks resisting sliding.

In UED, friction is the delay caused by incompatible ripple shells trying (and failing) to realign.

- When one body slides against another, their ripple fields overlap.
- If these fields are out of phase or misaligned in frequency or orientation, they resist smooth transition.
- This resistance is what we perceive as friction—not a contact force, but a temporal mismatch in ripple realignment.

Friction is alignment lag between bodies trying to coexist within the same ripple corridor.

And when that delay compounds—heat is released. Why? Because energy stored in ripple tension dissipates as consequence, not because anything rubs or burns by default.

Tension = Ripple Memory Storage

Tension in a rope. Strain in a beam. Stress in the body. What are they?

Not "forces inside objects." But stretched ripple memory, resisting collapse.

When a material is under tension:

- Its internal ripple configuration is extended beyond its equilibrium.
- The ripple shells are stretched, but still holding shape.
- This stored misalignment becomes potential for rapid correction—what we call tension energy.

Tension is the field remembering its rightful form, but not yet allowed to return to it.

It's not pulling. It's waiting.

And when released—like a snapping rope—the ripple field returns to equilibrium, unleashing its stored misalignment as visible effect.

Pressure = Ripple Re-alignment Demand Density

We call pressure a "force per unit area." But force never existed—so what is pressure really?

In UED, pressure is the density of ripple demand within a field zone.

- More ripple misalignments stacked in a region = greater pressure.
- The SVE field experiences compression tension—an urgency to rebalance.
- This is why pressure equalizes, not because molecules push, but because ripple shells collapse or stretch to accommodate field imbalance.

"Pressure is not a push. It's the buildup of unmet alignment trying to resolve itself."

That's why a balloon bursts—not because it couldn't handle the push, but because its ripple tension exceeded sustainable memory capacity.

Rewriting Mechanics in Ripple Terms

Classical Term	UED Reinterpretation
Friction	Alignment lag due to incompatible ripple shells
Tension	Stored ripple memory resisting collapse
Pressure	Ripple density imbalance within a confined field

"The universe never applies pressure. It waits for alignment debt to clear."

That's the key shift: UED sees all of this as debt settlement, not dynamic pushing.

These are consequence expressions, not initiators. There is no force acting—only a field trying to correct itself under the laws of structured memory and equilibrium.

3.7 There Are No Forces—Only Ripple Behaviors

Let's end the fantasy.

What we've called force—from Newton's apple to Maxwell's waves to Einstein's curvature—is not a truth of nature. It's a storyline invented to manage what we couldn't perceive.

In truth, there are no pushes, no pulls, and no actions at a distance. There is only structured vacuum energy (SVE)—rippling, compressing, re-aligning.

Every Force Dissolves into Field Consequence

What is a push? A reaction to overlapping ripple fields under misaligned torsion.

What is a pull? A surrender to ripple compression zones.

What is tension, thrust, or torque? Just field memories resisting, rebalancing, and restoring their historical configuration.

From electromagnetism to gravity, every so-called force is simply:

- A pressure differential in ripple density.
- A phase contradiction between two field regions.

- A delay in alignment handover.

There's no entity applying force. There's only the universe responding to ripple imbalance.

No Need for Laws When Behavior Is Sufficient

We didn't observe force—we invented it to save face when predictions failed.

- Newton said, "If it moves, something pushed it."
- Maxwell said, "If it radiates, fields did it."
- Einstein said, "If it curves, space bent it."

But none asked:

"What if everything you see is just how a field remembers itself?"

In UED, we stop asking what pushes what. We start asking:

- What caused this field misalignment?
- What tension is being released here?
- What alignment is attempting to restore itself?

This is not speculation. It's liberation from mechanical dogma.

Replace Force with Field Logic

Let's rewrite the classical lies:

Classical "Force"	UED Reality
Gravitational pull	Ripple compression collapse
Electromagnetic force	Field torsion and phase balancing
Contact force	Ripple interference resolving across boundaries
Normal/reaction force	Memory of equilibrium resisting ripple intrusion
Friction	Misaligned ripple shells delaying realignment

In short: Force was the excuse. Alignment was the truth.

So, What Moves Things Then?

Nothing moves. Nothing pushes. Nothing acts.

Things appear to move when ripple alignments transition in observable zones.

The object didn't go forward. The field updated. The tension didn't resist. The ripple refused to forget its last position.

What we witness as "force" is the choreography of ripple consequences, flowing like memory echoes across the structured vacuum.

Remember

"The universe never pushed anything. It only ever remembered where balance broke—and tried to fix it."

3.8 Closing: Force Was Our Word for Ignorance

For centuries, we called it a force.

Every time nature responded in ways we didn't understand—Every time something changed, moved, or resisted—We reached for that same word: force.

It became our comfort zone. Our shortcut. Our shield. But mostly—It became our excuse for not knowing.

Force Wasn't the Answer. It Was the Gap.

- "Why did it fall?" — Gravity.
- "Why did it slow down?" — Friction.
- "Why did it speed up?" — Thrust.
- "Why did it stick?" — Magnetism.
- "Why did it snap?" — Tension.

But these were descriptions, not explanations. We named behaviors. We didn't understand them.

Force was our placeholder. It sat where comprehension hadn't arrived yet.

It's not wrong to have placeholders. What's wrong is worshiping them as laws.

UED: No Forces. Just Field Truth.

Universal Energy Dynamics strips the illusion clean. No patchwork. No categories of push-pull. No zoo of equations.

Just one unified understanding:

Ripple alignment and misalignment drive all physical behavior.

Force isn't what the universe does. It's what we said it does—because we couldn't yet see how ripple memory and field tension interact.

In UED:

- Mass compresses ripple fields.
- Field resistance is consequence memory.
- Motion is realignment.
- Pressure is ripple density.
- "Forces" are ripple responses we dramatized into causal agents.

The field never forces. It corrects.

Goodbye, Force

Force wasn't real. It was a language artifact—a primitive tag we stuck onto behaviors we couldn't yet decode.

Now, with ripple logic clear, the word itself becomes obsolete.

Just like we outgrew "ether," "phlogiston," and "celestial spheres"—Force is next.

"The universe doesn't force anything. It just realigns what you disturbed."

Let that sentence echo through every textbook margin still teaching pushes and pulls as sacred law.

It's time to move forward. Not with force—But with clarity.

What We Just Shattered

Force is not fundamental. We exposed it as a conceptual crutch—a patch used whenever we failed to understand field behavior.

Gravity doesn't pull. It's not attraction at a distance, but ripple compression—objects align into lower ripple tension, not because they're "forced."

Magnetism is not a force. It's phase torsion. Poles are regions of ripple contradiction resolving themselves—not entities exerting invisible hands.

Electricity isn't flow. Charge is ripple imbalance. Current is not particle movement—it's the ripple field handing off alignment in sequence.

Friction isn't resistance from contact. It's a delay—misaligned ripple shells failing to re-sync instantly.

Tension and pressure aren't applied forces. They're ripple memories stretched and ripple density trying to settle.

Every classical "force" is a misread of ripple equilibrium. We weren't witnessing causality—we were watching consequence rebalancing.

The word 'force' itself is obsolete. It was never a substance, never a mechanism. Just the shadow cast by our ignorance of field intelligence.

"The universe never pushed. It never pulled. It waited for alignment to return."

Chapter 4: Time – The Greatest Fiction

Why the Clock Ticks Only Inside the Ripple Field

We never truly found time. We invented it—then got trapped inside our own invention.

From sundials casting shadows to atomic clocks pulsing with quantum regularity, we've tried to measure something that doesn't exist. What we call "time" is not a dimension. It's not a medium. It's not flowing. It's a consequence—a side effect of how the ripple field responds to imbalance.

In the absence of ripple delay, time has no meaning. Without an observer to experience realignment, there is no *before* or *after*—just the now constantly updating across the field.

Mainstream science made time sacred. Physics built equations that bowed before it. Relativity bent space to fit it. Thermodynamics crowned entropy as its arrow. Even philosophy surrendered to its tyranny.

But in the UED framework, time dies. Not because it's useless—but because it was never real to begin with.

You never moved through time. You waited through ripple restoration. And the ticking clock? That's just a synchronized echo of ripple recurrence—not a passage.

"You didn't age. Your alignment shifted. You didn't lose time. You changed memory."

Let's dismantle the greatest fiction we ever called a law.

4.1 The Birth of Time: From Sundials to Spacetime

Before we called it *time*, we called it *change*. A leaf turned yellow. The moon disappeared. Shadows grew longer. Early humans didn't measure time—they noticed difference.

And because our minds fear what they can't predict, we searched for rhythm in that difference. We marked seasons. We counted moons. We tracked sunrises and river floods. We built calendars not to measure reality—but to calm ourselves against its unpredictability.

Time wasn't discovered. It was carved into measurement by those who feared change.

Time as a Tool, Not a Truth

From sundials to water clocks, timekeeping evolved as a ritual of control. By the Industrial Age, time became currency—factories, railroads, and schools synchronized lives into hours, bells, and deadlines. But it was still externalized rhythm—not fundamental law.

Then came Einstein. His equations bent light and space under the weight of gravity and velocity. And in doing so, he gave time its greatest promotion: Not just a measure, but a dimension.

Spacetime was born—beautiful, brilliant, and misleading. It treated time as something we travel through. Something curved, stretched, and dilated.

But here's what got ignored:

Time was never a "thing." It was always an interpretation of ripple delay—how long it takes for the field to re-align, respond, or remember. Space doesn't need time. Ripples don't need clocks. Only *observers* do.

The Illusion Bakes In

Once time became a dimension, we stopped questioning it. We wrapped causality inside it. We framed existence around it. Birth, decay, motion, history—all became sequenced by the ticking of something that never ticked.

We assumed time must be real, Because we couldn't imagine experience without order.

But order is ripple alignment. Sequence is ripple transition. Memory is ripple tension. Time? Just a label on what consequence already knows how to do.

A Dangerous Assumption

Modern physics now orbits time. It divides equations into "t" terms. It predicts futures, explains pasts, and solves paradoxes by modifying time's behavior.

But UED doesn't need time.

Because once you understand the ripple delays between cause and effect, You realize: time is not the cause. It is the residue.

We don't move *through* time. We wait for the field to catch up.

4.2 Time Is Ripple Delay—Nothing More

Time has always worn a costume—dressed as movement, painted as flow, worshipped as a ticking god. But strip it of all metaphor and mythology, and you are left with a single, unavoidable truth:

Time is nothing more than ripple delay.

In the framework of Universal Energy Dynamics (UED), nothing flows. Not light. Not mass. Not space. Not even time. Everything realigns.

The Misinterpretation of Duration

Your watch doesn't track time. It tracks repetitive events—calibrated ripple realignments. An atomic clock vibrates in sync with energy transitions, not with "time itself." Seconds, minutes, hours? These are units we invented to quantify delay in the restoration of alignment between ripple fields.

The human brain—an intricate ripple detector—perceives change through these gaps. It doesn't observe time; it observes how long it takes for its surroundings, or internal state, to re-attune.

When you say "That took five seconds," what you're actually saying is:

"That ripple transition required five standard delay cycles as defined by my instruments."

Delay ≠ Flow

Mainstream physics still talks about time as a river—a forward flow no one can escape. But ripples don't flow. They ripple. They appear, compress, align, collapse, and realign.

Between cause and effect lies not a time interval, but a field restoration lag.

You don't walk through a hallway of time—you trigger a sequence of local ripple restorations. Your memory updates. Your observation shifts. Your body reacts. And the delay between these is what you call "time."

But in reality:

"There is no time—only how long a ripple takes to remember what it was."

Machines Count. The Mind Aligns.

A stopwatch can record 1.38 seconds. But your perception of that moment could feel like a blink or a lifetime, depending on your ripple sensitivity.

That's not subjective fuzziness. That is the truth revealing itself:

Time is a cognitive consequence, not a physical constant.

Where there is perfect alignment, time disappears. You call this "flow state," "deep presence," "timelessness." But really, it's just your ripple field syncing so well with its environment that no noticeable delay remains.

That's not mystical. That's physics—ripple physics.

4.3 Entropy Is Not an Arrow—It's a Record

In conventional physics, entropy is often described as the compass of time—the mythical arrow pointing from order to disorder. But UED throws this arrow back into the quiver.

Because entropy, when seen through ripple logic, doesn't mark time—it marks imbalance correction.

"Entropy doesn't prove time. It proves the memory of imbalance."

The Classical Mistake

Thermodynamics told us: entropy always increases. Therefore, time must go forward—never backward.

From melting ice to decaying stars, from broken cups to aged skin—disorder grows, and we mistake that growth for the passage of time. But that is like mistaking the smoke for the fire itself.

Disorder isn't the guide—it's the echo. Entropy is not time's engine. It's its receipt.

The UED Perspective: Entropy as Ripple Transition

In UED, every system—whether a gas cloud, a living cell, or a galaxy—is just a ripple configuration dancing in a structured vacuum energy (SVE) field.

When something appears to become disordered, it's not "moving forward in time." It is adjusting its ripple coherence to reduce the alignment pressure built up from earlier imbalance.

Entropy is the system's record of how far it is from equilibrium. Not a clock. Not a direction. Just a memory of what hasn't finished settling.

Equilibrium, Not Future

There is no forward. No later. No arrow.

There is only alignment recovery.

Entropy climbs not because we're headed to the future, but because ripple systems naturally relax under stress, spreading tension across more nodes until the burden becomes tolerable.

Disorder is not chaos. It is a strategy.

Just like a tense muscle softens. Just like a sound echoes before vanishing. Just like a mind forgets to heal.

Entropy is not destruction. It's decompression.

A World Rewritten

In UED, you don't "move forward" in time because entropy increases. Instead, you observe systems rewrite themselves to reduce ripple load.

Entropy does not define time. It confirms that the system once resisted balance—and is now correcting.

The arrow isn't forward. The arrow is toward coherence.

4.4 Aging, Decay, and Irreversibility

A wrinkle is not a sign that time has passed. It's a sign that your ripple field couldn't realign perfectly anymore.

We've been taught to believe that aging is the body's clock ticking away—day by day, cell by cell, second by second. But in UED, aging has nothing to do with the clock.

"You don't age because time passes. You age because your ripple field forgets its clarity."

Aging: Loss of Precision, Not Passage of Time

Every living being is a symphony of ripple alignment. Each cell, organ, thought, and emotion is held in place by the precision of ripple memory within the field.

In youth, the body's ripple field is sharp—coherent, adaptable, and responsive. But over time (or more accurately, over consequence), that field begins to strain under misalignment debt:

- Repeated stress = ripple noise
- Environmental toxins = ripple distortion
- Emotional trauma = deep misfire zones
- Poor nutrition = incoherent refresh cycles

Eventually, the system loses its re-alignment efficiency. The result? Wrinkles. Fatigue. Memory lapses. Structural decay. Not because time marched on—but because the field degraded in clarity.

Decay: Misalignment Buildup Beyond Recovery

In atoms, decay is framed as a probabilistic event—as if particles just get bored and disintegrate. But in UED, decay is ripple failure—a system's inability to hold its alignment due to accumulated imbalance.

It's not time-dependent, but tension-dependent.

A radioactive atom doesn't wait for time to pass. It waits for its threshold of ripple overload to break structure.

Decay isn't a countdown. It's a collapse—when realignment becomes mathematically unrecoverable within its field constraints.

Irreversibility: Memory Encoded as Resistance

Why can't we reverse aging? Why can't we "unburn" a log or "unsmash" a glass?

Because every ripple transition leaves a trace—a memory in the field that resists returning to a previous state.

This is not thermodynamic irreversibility. It's ripple imprinting.

Just like a path walked repeatedly becomes a trail, ripple transitions engrave themselves into the structured vacuum. Even if you "rebuild" the body, the field resists re-alignment to prior clarity unless every condition of coherence is restored.

Thus, irreversibility is not a metaphysical barrier—it is a field exhaustion phenomenon.

The UED Interpretation of Life and Loss

Aging is not time lost. It's precision lost.

Decay is not destiny. It's ripple fatigue.

And irreversibility is not a universal law. It is the price of encoded memory in systems that failed to hold their balance.

So if you ever ask, "Why do we grow old?" Don't blame time. Blame the ripple field—and its fading ability to remember what you once were.

4.5 There Is No Past or Future—Only Stored Ripples

We speak of time as if it stretches— Behind us, the past. Ahead of us, the future. But what if there's no line at all?

What if everything you've ever called "time" is just a misread ripple—the echo of an alignment your field once held, or one it anticipates holding?

"The past isn't behind you. It's still inside your ripple tension."

Memory Is Not a Timeline—It's Resonance

In UED, memory is not a data archive. It is a resonance state—the ripple field preserving certain alignments that once occurred.

You don't remember the past. You re-align to the way you once held the world.

The smell of a home-cooked meal doesn't take you "back in time." It re-activates a field configuration that still resides within your structure.

This is not recall—it's ripple reformation.

There is no "past" playing out somewhere else. It is you—your current ripple field—that replays the alignment.

Imagination Is Predictive Alignment, Not Future Access

When you imagine the future, you're not seeing ahead. You are running field simulations—your ripple field trying out possible alignments based on its current state.

That's why imagination is often emotionally charged. Because it isn't conceptual—it's energetic.

A child doesn't "visualize" flying. Their ripple field actually creates the sensory impression of it—eyes closed, arms wide, field responding.

This is not trivial. It means "the future" doesn't exist either. It's not waiting to arrive. It's just a field configuration not yet triggered.

Why Time Travel Is Impossible

Time travel assumes that time is a dimension you can traverse—like walking through rooms in a corridor of events. But UED sees no such structure.

There's no timeline. There's only consequence memory—and ripple patterns that can't be recreated perfectly, because their surrounding field has already evolved.

You can't return to yesterday, because the field tensions that supported that alignment are no longer available.

You can't skip to tomorrow, because it doesn't exist until ripple equilibrium creates it.

Past, Future, and the Truth of Now

So where is the past? In ripple memory.

Where is the future? In ripple potential.

What exists? Only the now—a constantly refreshed alignment stage, shaped by memory residues and field tendencies.

We are not passengers on a time train. We are conductors of ripple states—reassembling fragments of resonance into coherent experience.

"You are not older than yesterday. You are just holding more ripple scars. You are not closer to tomorrow. You are just aligning with its possibility."

4.6 Clocks Don't Measure Time—They Display Regular Ripple Shifts

The greatest illusion of time is not found in philosophy— It's sitting on your wrist. Or blinking from a wall. Ticking... as if the universe has a metronome.

But clocks don't measure time. They manifest ripple regularity—a controlled alignment loop mistaken as "time."

"Clocks don't tick through time. They align through repetition."

What Clocks Really Do: Ripple Synchronization

Every timekeeping device—from a sundial to an atomic clock—is a tuned ripple loop. It's a system whose internal ripple state shifts with precision, then returns to its baseline on cue.

- A pendulum clock: the gravitational ripple field induces periodic realignment of the bob.
- A quartz watch: electric ripple tension in the crystal vibrates consistently.
- An atomic clock: microwave pulses sync with energy level shifts (ripple shells) in cesium atoms.

Each of these systems is not tracking an external thing called time. They are performing field routines under controlled conditions.

What they actually measure is consistency of ripple re-alignment.

Why Atomic Clocks Don't Prove Time

Scientists call atomic clocks the "gold standard" of precision. But what are they really measuring?

Cesium atoms don't "know time." They simply flip their ripple shells at a constant energetic threshold under lab-isolated conditions.

We interpreted these ripple shifts as "9,192,631,770 oscillations = 1 second."

But that's not a discovery of time. That's a definition of ripple rhythm.

Change the field conditions—temperature, pressure, gravity—and your "second" stretches or shrinks.

So what did we really calibrate? Not time. Field behavior.

UED Replaces Timekeeping with Field Logic

In UED, there's no need to invoke "universal time" or "space-time curvature" to explain dilation or synchronization.

It all boils down to ripple phase logic:

- Ripple delay = perceived duration.
- Ripple tension = drift or lag.
- Field compression = perceived speed-up.
- Ripple decoherence = time loss or blur.

The ticking of a clock is not universal rhythm—it's a repeated ripple decision under steady conditions.

Once those conditions shift, so does your measurement.

The Real Question: Why Do We Trust the Clock?

Because it feels consistent. But so does the Earth's rotation. So does the beat of a drum. So does your own heartbeat—until it doesn't.

The clock comforts us with the illusion that something is stable. But UED reveals: it is only field choreography, and nothing more.

"A clock doesn't tick for the universe. It ticks because the ripple field repeats its steps... and we call that time."

4.7 Closing: Why Time Is the Most Elegant Lie We Believed

Of all illusions we've embraced as truth— Time was the most elegant. Not because it was proven. But because it was felt.

The heartbeat, the sunrise, the wrinkle, the decay...They all whispered: *"Something is passing."* So we built a word. We called it time.

"The universe never kept time. It just held its breath until you noticed."

There Is No Time—Only Consequence Delay

Let's recap the real mechanics according to UED:

- Events don't occur in time. They occur through ripple misalignment reaching resolution.

- Duration doesn't exist as a thing. It's your field's struggle to re-align—what we call "waiting."
- Sequences aren't measured by clocks. They are sensed as field memory activating in phase.

In short: Time is not flowing. It is unfolding ripple pressure.

Time Was Never a Dimension—It Was Always an Interpretation

Einstein taught us to bend time. We applauded. But we never asked what time was in the first place.

If space bends and light slows, does "time" change? Or is it just that your ripple re-alignment shifted?

Every so-called dilation, compression, aging effect—These are ripple behaviors under energetic strain—not temporal dynamics.

We didn't detect time. We witnessed field fatigue, delayed coherence, alignment lag.

And because we couldn't name it— We called it "time."

Without Ripple Misalignment, Time Would Not Exist

In perfect equilibrium, nothing "changes." No new ripple, no delay. No misalignment = no need to track realignment = no perceived time.

This is the true definition of timelessness: Not eternal stillness, but complete ripple coherence.

The Grand Lie, Beautifully Told

Time helped us organize. It gave meaning to memory. It let us schedule, forecast, anticipate, and mourn.

But like all grand myths, its beauty was in its utility, not its truth.

Time is not a force, dimension, or cosmic river. It is a mental calibration to field consequence. When the ripple clears, Time dissolves. And all that remains is now.

"Time never pushed the universe forward. The universe only pushed until something aligned—and you called it a second."

What We Just Shattered

Time is not a thing. It is a delay. We shattered the myth that time exists independently of you. Time is born out of ripple misalignments and their resolution. No ripple delay, no time.

Clocks don't measure time—they display repetition. Atomic precision doesn't prove time's existence. It only proves that ripple systems can be tuned to align predictably—and we mistake that rhythm for something fundamental.

Entropy isn't an arrow—it's memory tension. We exposed entropy as ripple history trying to balance itself—not a law pushing the universe forward.

Aging and decay are not effects of time. They are signs that ripple coherence is breaking down. You don't grow old because time flows—you grow old because memory alignment fades.

There is no past. There is no future. Only stored ripple tensions and predictive realignments. Memory and imagination simulate a timeline—but there is no "line" to travel.

Time was never real. It was only useful. Like force, particles, and motion, time was a placeholder for what we couldn't yet explain. Now, with UED, we no longer need that placeholder.

"You didn't live in time. You lived in ripple delay. And you called it a life."

Chapter 5: Spacetime – A Beautiful Lie

Why the Universe Doesn't Bend—It Aligns

There's a reason "spacetime" became one of the most revered concepts in science. It was elegant. Powerful. Visual. It made gravity look graceful and mystery sound intelligent.

Einstein didn't just reshape physics—he gave it a metaphor. He said: *Mass tells space how to curve, and space tells mass how to move.* And with that poetic symmetry, the world fell in love with a lie.

Because nothing ever curved. Nothing ever moved. And "spacetime" was never anything more than a mathematical seat cushion for gravity—an illusion padded to comfort us when we couldn't explain what actually happens.

In Universal Energy Dynamics (UED), there is no fabric. There is no stretch. No warp. No bend. There is only ripple compression. Pressure. Field equilibrium. And what looks like a curve... is just your alignment trying to keep up.

"You didn't fall into gravity. You surrendered to compression."

This chapter isn't here to disrespect Einstein. It's here to complete him. To show where geometry distracted us from deeper causality. To replace spacetime curvature with ripple realignment. And to remind the world that:

Prediction isn't truth. Visualization isn't reality. And beauty doesn't make it real.

Let's tear the fabric.

5.1 The Seduction of Geometry

"We didn't find curvature—we imagined it to escape force."

For over a century, the scientific community has been hypnotized by one of the most beautiful ideas ever introduced in physics: the curvature of spacetime.

Einstein's stroke of genius wasn't in discovering some hidden truth—it was in framing gravity as a geometrical story rather than a brute-force mechanism. Instead of action at a distance, he gave us a universe of smooth deformations. Instead of invisible pulls, he offered warped paths and geodesics. Gravity didn't need a force anymore—it just needed a hill to roll down.

This was revolutionary. It was elegant. And it was seductive. But it was still a model. A metaphor. A visual compromise dressed up as causality.

Because despite all its predictive power, spacetime curvature was never something we observed. We didn't see space bending. We didn't measure time stretching. We saw outcomes—delays, accelerations, deviations—and reverse-engineered a geometry to fit.

Curvature became a crutch, a placeholder for an unknown mechanism—just like "force" had been before it.

But what if there is no bend? What if there is no fabric to stretch?

UED throws away the crutch.

It doesn't need imaginary warps to explain attraction. It doesn't require mathematical patches like tensors and coordinate frames. Instead, it proposes a simpler, more foundational truth:

Gravity is not curvature. It's compression. Mass doesn't warp spacetime. It compresses ripple fields. And what we see as acceleration is just alignment shifting toward a zone of lower ripple tension.

Einstein's metaphor worked because we needed it. But like any metaphor, it only took us so far before it began hiding the truth it was meant to reveal.

Spacetime curvature didn't liberate us from Newton—it softened the same idea with math.

And now, a century later, it's time to stop being seduced by geometry...and start listening to the ripple beneath it.

5.2 Why Space Is Not a Fabric

"Space doesn't hold you. It remembers you."

For decades, we've described space as if it were something. A *thing.* A fabric. A grid. A container. A medium upon which reality is placed—stretched, bent, curved, torn.

We've imagined it like rubber. We've animated it like cloth. We've even spoken of "ripples in spacetime" as though it were a trampoline hosting galaxies like bowling balls. But here's the uncomfortable truth mainstream science won't admit:

No one has ever found the fabric. No evidence. No particles. No properties. Because it doesn't exist.

Spacetime—as a substance—is a poetic myth. It's a placeholder. A well-behaved illusion.

Under Universal Energy Dynamics (UED), we discard the metaphor. Space is not a stage. Not an arena. Not a passive background. It is not *there* waiting for things to move through it.

Space is not empty. It is active. It is not a fabric—it is a ripple field in continuous tension, memory, and adjustment.

In UED, space is nothing but structured ripple potential within the Structured Vacuum Energy (SVE). Every "point" in space is not a dot in a coordinate system—it is a knot of potential alignment, ready to host a ripple manifestation. It remembers previous alignments, stores tension patterns, and shapes the very conditions for what can appear or act within it.

When you say something exists "here" or "there," you're not locating it in geometry—you're describing which ripple pattern has achieved equilibrium at that zone of SVE. That's why we say:

Location is not position—it's permission. A node is "somewhere" because the field allowed it to be—not because there's an address labeled X, Y, Z.

In this view:

- Movement is ripple transition.
- Proximity is coherence potential.
- Distance is ripple misalignment.

So why did we ever call space a "fabric"? Because we needed something to explain how light bent, how gravity acted, how time dilated. We turned abstraction into assumption. We mistook model for matter. And in doing so, we started sewing threads into a vacuum that never needed stitching.

It's time to drop the loom.

"Space is not what wraps around the universe. It is how the universe remembers how to ripple."

5.3 The Illusion of Curved Paths

"The universe didn't curve the road. It changed the conditions of your ripple."

Take a moment and think of every cosmic diagram you've seen since school: Planets arcing elegantly around the sun, comets tracing ellipses, light bending near massive stars—all drawn as graceful curves on a stretched grid of space.

We were told those curves were real. That planets roll through spacetime like marbles in a bowl. That gravity is geometry. That light has no choice but to follow the bent road.

But what if there is no road? What if there is no curve?

UED exposes the lie beneath the arc.

In Universal Energy Dynamics, orbits and bends aren't paths—they're ripple alignments. The planets are not "moving through curved space"—they are stabilizing in the ripple pressure gradient of the surrounding field.

Let's decode it:

- A planet's orbit is the most efficient ripple memory path within a compressed field zone. Not because it's being pulled. Not because space is warped. But because the ripple field offers minimal resistance along that alignment loop.
- Light near a star doesn't bend because space is curved—it re-routes because the ripple field has been densified by mass compression. The light doesn't change direction—it emerges along the new minimum-tension pathway allowed by the field.

Think of it like this: If you shout in a canyon, your voice bounces and redirects—not because you changed direction, but because the environment shapes the ripple's continuation.

Likewise, a beam of light isn't "dragged" by curved spacetime—it is redirected by field distortion, just like sound waves bending through temperature gradients or seismic waves curving through Earth's layered interior.

What we mistook for a bent road... was actually a rerouted signal.

And the curve? It wasn't in space. It was in our assumption that motion requires a path.

"A planet doesn't orbit. It resonates. A beam of light doesn't detour. It re-emerges in allowed alignment."

So the next time you look at an orbit diagram or an Einsteinian grid warped around a sphere—don't be fooled by the elegance. It's not reality. It's a stylized metaphor drawn to comfort your geometry-loving brain.

Reality doesn't curve. It compresses. Alignment shifts—not because of invisible warps—but because of ripple pressure logic trying to restore equilibrium.

5.4 Time Was the Patch—Not the Explanation

"We stretched time to save gravity—but ended up losing both."

General Relativity (GR) was a masterpiece—not of truth, but of mathematical repair.

When Newton's force-based gravity couldn't explain anomalies like Mercury's orbit, or the bending of light near stars, Einstein came along and did something bold: He didn't just fix gravity. He rewrote the language in which it was spoken.

He merged space with time into a four-dimensional stage called spacetime—where mass didn't pull but bent this stage, and motion wasn't force-driven but geodesic—an illusion of the shortest path in curved geometry.

But here's what no one paused to ask:

What if the problem was not space or force...but the very idea of "time" itself?

GR didn't eliminate force. It just renamed it.

It traded Newton's "pull" for a "path", and time became the flexible thread that let the theory work—not because it was real, but because the math needed it.

Einstein turned time from an independent clock into a curved coordinate, but that didn't make it fundamental—it made it a patch.

- Why did clocks slow down near massive objects? GR said: Because time dilates.
- Why did light shift frequency when climbing out of a gravity well? GR said: Because time ticks slower down there.

But UED says something different—and much simpler: Time is not flowing. The ripple field is compressed. Delays increase. Alignment slows. Memory stretches.

No need to warp a fourth dimension. No need to curve imaginary grids. What you're witnessing is ripple tension—not temporal distortion.

In UED, gravity doesn't bend time. It densifies memory.

What appears as time dilation is actually ripple delay—a slowing of the field's capacity to re-align and propagate influence. What looks like gravitational redshift is a field attempting to maintain coherence under pressure.

This changes everything.

- No more clocks embedded in fabric.
- No more light following bent geodesics.
- No more mystery about time "flowing slower" near black holes.

Because there is no "flow" at all. There's only how long the ripple takes to remember what it was.

So when we say "spacetime"—what we really mean is:

A placeholder to rescue Newton with elegance. A patch to stitch a force-shaped hole with curvature. A compromise, not a cause.

And as with all patches in science—once you see the thread... the whole tapestry begins to unravel.

5.5 Gravitational Lensing Reimagined

"Light wasn't bent. It was rerouted by ripple surrender."

Gravitational lensing is one of the most cherished "proofs" of General Relativity. When light from a distant star curves around a massive object like a galaxy or black hole and reaches us at multiple angles, mainstream science celebrates:

"Spacetime curved. Light obeyed. Theory confirmed."

But in truth, nothing curved. Nothing obeyed. And light didn't even *move*.

What Really Happened?

According to Universal Energy Dynamics (UED), light is not a stream of particles or a wave traveling through space—it is a consequence of ripple alignment, a perception of ripple-phase harmony between source, medium, and observer.

So when you observe gravitational lensing, what you are seeing is not:

- A photon flying through curved spacetime
  But instead:
- A ripple realignment occurring along the most memory-compatible path.

Let's break it down:

Mass compresses ripple fields.

A dense object like a star or galaxy contracts the surrounding structured vacuum energy (SVE), increasing ripple pressure locally.

Ripple memory guides propagation.

UED proposes that ripple propagation doesn't follow straight lines in a void—it follows the path of least alignment resistance, akin to how water flows through terrain not because the terrain is "curved" but because the path is energetically favorable.

So the light doesn't bend—it re-emerges in your field of perception from a different alignment pathway, detoured through compressed ripple zones.

What does this mean?

- The observer perceives multiple images not because light split or bent—but because multiple ripple routes are coherently valid under the present field tension.
- The "lens" is not curving space—it's distorting ripple

memory so that incoming alignment signals reorganize themselves around it.

You're not seeing the distant galaxy directly. You're seeing which ripple paths could still make it through to your awareness after navigating the compressed ripple landscape.

This collapses the need for spacetime curvature entirely.

Implication:

Gravitational lensing does not validate General Relativity. It exposes the weakness of interpreting light as motion and space as geometry.

Instead, it affirms:

- Light is ripple alignment.
- Gravity is ripple compression.
- Perception is a map of memory-tension compromise.

So let's rewrite the phenomenon once and for all:

"Light wasn't bent. It was rerouted by ripple surrender." And what you saw—wasn't the object. It was the alignment that remained.

5.6 Why Spacetime "Works"—But Fails

"Success in prediction does not mean success in truth."

General Relativity (GR) is undeniably successful in *predicting outcomes*. It has guided GPS corrections, explained Mercury's perihelion shift, and forecasted black hole behavior with remarkable numerical precision.

But prediction is not explanation. And that's where the lie begins.

GR's Success Is Mathematical, Not Foundational

Einstein's spacetime curvature model predicts what happens, but it never explains:

- Why curvature forms at all.
- Why mass bends space.
- What space is made of.
- Why light should follow a curved path.

It bypasses ontological clarity and replaces it with elegant distortion: A four-dimensional *fabric* with no fabric. A mathematical surface that nobody can touch, measure, or define as a medium.

This is why physicists celebrate equations but quietly dodge the questions:

"What exactly is curving?" "What is space, really?" "Is time stretching... or are we just lagging?"

UED Exposes the Cracks

In Universal Energy Dynamics (UED), there is no fabric—only ripple alignment across a structured vacuum field (SVE). Mass doesn't bend geometry—it compresses local ripple zones, raising alignment pressure. Light doesn't ride geodesics—it re-emerges along memory-favorable ripple conditions.

So yes, GR "works"—but only because it's numerically tuned to outcome, not truth.

It's like plotting the shadow of a mountain: you can trace it perfectly, but it still doesn't tell you what the mountain is made of.

The Trap of Predictive Success

Mainstream physics clings to GR not because it explains reality, but because it avoids chaos with calculable shortcuts.

Here's the uncomfortable truth:

GR doesn't know what a field is. GR doesn't explain what mass is. GR doesn't tell you what time really is.

It just creates a geometry of effects around undefined causes.

UED's Advantage: One Field, Many Consequences

UED does not need spacetime. It needs only ripple alignment logic. From light to inertia, gravity to charge—every phenomenon emerges from how energy disturbances (ripples) interact with the structured vacuum.

- No dualities
- No geometry patches
- No placeholder dimensions

Just one ontological field giving rise to many perceptual consequences.

So why did spacetime survive for a century?

Because it *predicted* what the eye could see. But it never described why the ripple behaves as it does—because it never knew the ripple existed.

"Spacetime worked like a prosthetic leg—it let us walk the map of physics. But now we've grown new legs: ripple intelligence that explains what was always there."

5.7 Closing: The End of Geometry Worship

"The universe doesn't stretch or curve. It recalibrates alignment—and we call it shape."

We fell in love with geometry—and forgot to ask if it was even real.

Spacetime gave us elegant metaphors: curves, warps, bends, and wells. It let us imagine gravity as a visual playground, complete with grids, dips, and balls rolling on stretched sheets.

But here's the cost of that metaphor: We stopped seeking cause, and started simulating effect.

Visual Thinking Trapped Physical Reasoning

Einstein's genius was poetic: he turned physics into a painting. But like all great art, it's open to interpretation— and misinterpretation became our doctrine.

- We pictured space as a fabric, without ever identifying the material.

- We imagined time as a dimension, without proving it exists independently.
- We treated curvature as action, forgetting that it's just a *reaction* to something we still can't define.

Geometry didn't fail. We just mistook it for truth.

The Simulated World Isn't the Real One

Spacetime worked well as a simulation environment— a framework that predicted trajectories, paths, and orbits.

But simulation is not reality. It's a shortcut through consequence, not an understanding of origin.

UED challenges that illusion head-on: It says: "Drop the canvas. Study the ripple." What we call "shape," "structure," or "motion" is nothing but realignment within ripple fields—not forms placed in space, but behaviors of space itself.

UED: Beyond Geometry, Into Field Logic

The universe is not a 4D arena with lines and loops—it's an intelligent ripple choreography.

- Mass compresses alignment fields—not space.
- Light detours through ripple tension—not curved spacetime.
- Time unfolds as delay—not as a flowing coordinate.

Everything you thought was geometry... was just ripple trying to restore memory. So let's say goodbye—not to Einstein's brilliance, but to our addiction to the metaphor.

"Physics must not worship drawings. It must study disturbance, recovery, and memory."

Because the universe doesn't sketch. It self-aligns. And what you see as shape, curve, or force was never the structure—it was the signal of balance being regained.

What We Just Shattered

"Space is not a fabric—it's a ripple field." We dismantled the romantic illusion of spacetime as a stretchable canvas. There is no physical medium to bend, stretch, or ripple. What you call "space" is an active memory system—structured vacuum energy in ripple alignment.

"Curves do not cause motion—ripple compression does." Planetary orbits, light detours, and gravitational arcs aren't results of bending space. They are the result of ripple gradients and pressure zones rebalancing themselves.

"Light was never bent—it surrendered to ripple tension." We reinterpreted gravitational lensing not as light bending through curved space, but as ripple compliance with field pressure zones. Light appears redirected because ripple continuity shifts—not because space was deformed.

"Time didn't fix gravity—it hid its cause." Einstein added time as a patch for Newton's framework. But in doing so, physics began worshipping equations over consequences. UED replaces time with ripple delay and compression memory—removing the need for a temporal dimension altogether.

"Geometry is simulation, not truth." Spacetime gave us good predictions, but no foundational understanding. A simulation is only as good as the logic it hides. Geometry only reflects ripple outcomes—not their cause.

"There is no curved spacetime—only ripple tension zones." When ripple fields compress, alignment paths distort. You see curvature. But it's not a shape you can stretch—it's a **field you misunderstood.**

Core Conclusion: "You didn't watch space bend. You watched the ripple adjust its memory—and mistook it for a map."

Interlude – Cracks in the Temple

They taught you particles.
You imagined dots.
They taught you motion.
You imagined travel.
They taught you force.
You imagined power.
They taught you time.
You imagined flow.
They taught you spacetime.
You imagined elegance.

And somewhere along the way, you forgot to ask the one question that undoes it all: "What if none of it is real—just a pattern of consequences decoded by alignment?"

You were never living in a machine built from parts. You were living inside a choreography of ripple tensions—structured by equilibrium, memory, and surrender.

Part I shattered the scaffolding. It dismantled the illusion of a universe made of things, moving through space, pushed by forces, tracked by time, wrapped in curvature.

It showed that what we called laws were often just translations of ignorance, encoded in math to hide our discomfort with uncertainty.

So what now?

Now that the temple has cracks—do you run back to comfort?

Or do you step through the ruins and see what was hiding behind the altar?

Because behind it...There is no particle. There is no field. There is no equation.

There is only the consequence—rippling. There is only the memory—realigning. There is only the observer—realizing.

Part II: Rebuilding the Real

Chapter 6: Universal Energy Dynamics

A Reality Framework

Why Consequence Is the Only Cause That Matters

We've broken the illusion of particles. We've dismantled the myth of motion. We've exposed force as compensation and time as delay. Now, it's time to ask: If none of these were real, then what is?

Welcome to **Universal Energy Dynamics (UED)**—not a theory, not a force, not a formula. A *framework*. One that doesn't try to patch reality—it reads it. A model that doesn't begin with matter or motion, but with alignment, resistance, and the memory of imbalance trying to heal.

In UED, you don't need particles, forces, or a timeline. You need only one field—**Structured Vacuum Energy (SVE)**—and its constant dance of ripple emission, alignment, delay, and release.

Everything you've ever touched, felt, feared, or fought for was not caused. It was *consequential*

This chapter doesn't explain the universe. It shows you what it was doing while we were busy explaining.

6.1 The Premise of UED: Energy Is Not Acting—It's Remembering

We've long assumed energy acts—that it pushes, propels, and produces. But this assumption was never proven. It was inherited—from classical physics, from the industrial age mindset, from our obsession with action. Universal Energy Dynamics begins by flipping that script entirely:

Energy is not acting. It is remembering.

Everything you observe—a flame flickering, a planet revolving, a heartbeat pulsing—is not a fresh event. It is a consequence unfolding along remembered alignments of ripple fields. Nothing is happening for the first time. It is all echo—a sequence of memory realignments seeking equilibrium.

Where classical science begins with cause, UED begins with consequence.

Cause is speculative: it presumes a trigger. Consequence is visible: it insists through outcome. This is why UED does not ask, "What started it?" It asks, "What insists through it?"

Imagine ripples in a pond. One drop. One memory. But the patterns don't vanish—they realign, collide, delay, imprint. And when another ripple appears, it doesn't act. It responds. It is shaped by the memory already present.

This is not poetic metaphor—it is the foundation of the observable universe.

Everything you think of as "event" is ripple behavior collapsing into observable consequence. You didn't make the decision—your alignment did. The rock didn't fall—its ripple memory reentered equilibrium. The sun didn't burn—its ripple field refused compression delay.

"You are not the cause. You are the echo."

In UED, this isn't just philosophical. It's mechanical.

Reality is not created. It's recalled. And you, the observer, are not a creator of reality—you're a ripple interpreter, decoding the memories of universal alignment.

This section resets everything we believe about agency, force, and interaction. If energy is remembering, then all systems—biological, physical, emotional, or cosmic—are manifestations of alignment trying to repeat, reconcile, or release.

So the question is no longer, "What caused it?" The question is, *"What memory is this moment honoring?"*

6.2 Structured Vacuum Energy (SVE): The Hidden Ground

Forget what you've been told about empty space. There is no such thing as *empty*. The vacuum is not void. It is structure waiting to respond.

Structured Vacuum Energy (SVE) is the core substrate of Universal Energy Dynamics. It is not a fabric. It is not a quantum foam. It is not spacetime.

It is perfect equilibrium—a field of undisturbed potential where nothing moves, but everything is poised for consequence. Not because it's static. But because it's balanced to the point of invisibility.

SVE is tensionless, but not lifeless. It is ready, but not reactive until disturbed. You might say it is the *cosmic memory foam* of reality—holding no form, but retaining the exact readiness to form the moment anything asks it to.

This is where everything begins—not with matter, motion, or force, but with a breach in this equilibrium. That breach, that misalignment, is what UED calls the first ripple. And every ripple after is not new—it is a compensation, an echo, a return attempt.

In this view:

- A star forms? That's SVE collapsing into local compression.
- A photon appears? That's ripple tension breaking threshold alignment.
- A thought emerges? That's conscious ripple reconfiguration in a biological shell.

All ripple phenomena—particles, waves, light, fields—are expressions of the SVE under pressure.

SVE itself cannot be detected directly, because detection requires disturbance. The moment you try to "see" it, you've already created a ripple. And what you see is not SVE—but the consequence of its disruption.

"Nothing is empty. Everything is poised."

This insight changes everything. The background is not neutral—it is the origin of consequence. And the universe is not full of particles—it is full of ripples trying to restore what SVE once held still.

So when you witness a galaxy spin, or an electron jump, or even a heartbeat beat—don't look for the object. Look for the ripple. Because the object is never the truth. The truth is the field behind it, holding perfect silence until reality breaks the surface.

6.3 Ripple Emission: How Energy 'Begins'

Creation never needed a spark. It needed a misalignment.

Energy does not begin by being added into the universe like fuel into a fire. It emerges as a result—a response to imbalance in what was once fully balanced.

In Universal Energy Dynamics (UED), this is known as Ripple Emission: The birth of every observable phenomenon, not through injection of force or material, but through disturbance in the Structured Vacuum Energy (SVE).

Energy is not injected—it emerges

In legacy models, we speak of energy being transferred, conserved, or applied.

But UED proposes something deeper:

- Energy is not a thing being moved from one location to another.
- It is a pattern that appears when equilibrium is disrupted.

- Ripple emission is not a push. It is a reminder that something has deviated.

The moment SVE is no longer perfectly silent, a ripple forms—an echo of that disturbance.

Any disturbance generates a ripple

Ripple emission is inevitable wherever:

- A local event breaks symmetry.
- An intent arises from a conscious observer.
- A force tries to act (which is already a consequence, not a cause).

Whether it's the fusion reaction in a star or a thought in a mind, any interference with the stable equilibrium of SVE emits ripples.

These ripples are not always neat. They vary:

- Directional ripples form under clean asymmetry (e.g., light beams).
- Coherent ripples occur in resonant fields (e.g., laser, atomic structures).
- Turbulent ripples form when misalignment is chaotic (e.g., explosions, emotional surges).

In all cases, what we perceive as energy is simply the intensity and quality of ripple emission from the SVE field.

Ripple strength = degree of misalignment

Not all ripples are born equal.

The strength of any ripple is determined by:

How far the local field has deviated from its equilibrium.

- A small misalignment: weak ripple.
- A large distortion: intense, high-energy ripple.

- A sustained disturbance: ripple node (what we falsely call particles).

Thus, energy is not a quantity—it is a signal of imbalance.

A massive object? Just a strong ripple node holding persistent tension. A flash of gamma radiation? A sharp, sudden misalignment discharging violently. A heartbeat? Repeated controlled ripple emissions within a bio-alignment field.

Creation is the misalignment of silence

Let that phrase echo through the rest of this book.

"Creation is the misalignment of silence."

Because nothing was ever "created" as a standalone event. Everything you call existence is merely SVE breaking into ripple—and ripple being interpreted as energy, motion, mass, or matter.

It all begins with a breach in balance. Not with an origin story. Not with a bang. But with a whisper that broke the stillness.

6.4 Alignment: The Currency of Reality

"Existence is ripple alignment sustained."

In a universe that neither builds nor breaks—but remembers and restores—the most valuable phenomenon is not matter or energy.

It is alignment.

Wherever ripple patterns stabilize, we get the illusion of something real—something that "exists." But what we truly observe are zones of coherence, held together by the natural pull of equilibrium. These are not solid things. They are consequences locked in pattern.

Reality = zones where ripple patterns stabilize (ripple nodes)

Reality doesn't "exist" in a blanket sense. It is selective emergence—localized clarity within a sea of possibilities.

In UED, these localized regions are called ripple nodes:

- They are not particles.
- They are not structures.
- They are knots of coherent ripple memory that sustain a pattern in SVE.

Just like a whirlpool appears stable in a flowing river, ripple nodes are stable disturbances—not static entities.

This is what we've always mistaken as the "material world."

Mass, charge, field: all are alignment profiles

Let's tear down the textbook definitions.

- Mass is not an intrinsic quality. It is the resistance of a ripple node to re-align.
- Charge is not a property. It is a phase offset in the alignment rhythm of a field.
- Electric, magnetic, gravitational fields? They are maps of ripple tension gradients, not separate forces.

Everything we observe—weight, attraction, repulsion, interaction—is a result of how well ripples align with one another in time and space.

Alignment is the information. Alignment is the interaction. Alignment is the only reason anything seems to "persist."

Realignment pressure = what we call 'forces'

What we commonly call a force—push, pull, gravity, magnetism—is nothing more than the pressure to restore or disrupt alignment.

There is no need for mysterious "interactions at a distance." There is only field memory trying to return to equilibrium.

Examples:

- A falling object is not "attracted" to Earth. → It is realigning with Earth's ripple compression zone.

- A magnetic repulsion is not two poles pushing away. → It is a phase contradiction between ripple fields unable to merge cleanly.

In every case, force is not the cause. Realignment is the consequence.

"Existence is ripple alignment sustained."

Nothing exists in a permanent sense. Everything is in flux—but some patterns hold together longer.

And why do they hold? Because their alignment is locally stable enough to resist collapse.

A rock. A star. A thought. A habit. All are temporarily sustained ripple alignments.

We do not live in a physical world. We live in a coherent ripple performance, constantly rewriting itself as it balances, corrects, and remembers.

Existence = memory + rhythm + equilibrium. Not matter. Not mass. Not motion. Just alignment—and the stories we build around it.

6.5 Delay: The True Origin of Time

"The slower the ripple realigns, the longer the illusion of time."

Time, as we've always defined it, was a convenient scaffold: A way to track events. To sequence them. To make sense of cause and effect.

But what if time isn't flowing?

What if time is merely lag—the ripple field's resistance to immediate re-alignment?

That's what Universal Energy Dynamics (UED) proposes: Time is not a dimension. It is a delay profile.

Ripple delay = perceived duration

Every field in the universe strives toward equilibrium. But not all ripples realign instantly.

Some adjust easily. Others take longer to correct. That delay—the time it takes for a ripple to "remember" its optimal state—is what we perceive as time.

You're not watching seconds tick. You're feeling the delay of consequence being handed off across ripple zones.

A fast alignment appears "instant." A slow one feels like "duration." This is the only reality behind our clocks, calendars, and countdowns.

Time is born from ripple handover lag—not an absolute

There is no universal time—no cosmic metronome keeping rhythm across galaxies.

What we experience instead is the handover lag: The latency between one field zone aligning and the next one responding.

- When the lag is short: the world feels "fast."
- When the lag stretches: we feel "stuck in time."

Even in physics, this appears as wave delay, signal latency, or phase difference.

But it was never about time moving forward.

It was always alignment taking its own pace to resolve conflict.

Alignment rate ≠ speed of motion, but ease of consequence transfer

In classical terms, we talk about velocity, momentum, or acceleration.

But in UED:

- Speed doesn't describe how fast an object moves.
- It describes how easily ripple consequences are transferred across space.

Think of a baton relay—not the runner's legs, but how quickly the baton changes hands.

That is the true speed: How smoothly alignment is passed without distortion or interference.

Motion is not a property. It's a symptom. Alignment transfer is the mechanism.

"The slower the ripple realigns, the longer the illusion of time."

Imagine a stretched rubber sheet with a ripple disturbance.

- If it snaps back quickly: no time is felt.
- If it wobbles slowly: we "feel" its passage.

That wobble, that lingering misalignment, is where your experience of time is born.

Time doesn't tick. It waits—for realignment to finish.

We age because our ripple field takes longer to recover. We feel pain because the alignment doesn't reset instantly. We grieve, we heal, we remember—because our field delays let the consequence linger.

There was never a ticking universe. Only a patient field, resolving at its own rhythm.

6.6 Memory: Why the Universe Doesn't Forget

"Reality is what the field refuses to forget."

If ripple delay gives us the illusion of *time*, Then ripple memory gives us the illusion of *persistence*.

What we call a "thing" isn't an object—it's a pattern remembered by the field. And what we call a "law" isn't a rule—it's a recurring alignment outcome from a remembered past.

The universe doesn't evolve through commands or scripts. It evolves through repetition—repeating ripple alignments encoded by consequence.

Ripple paths imprint field memory

Every ripple, once emitted, leaves an imprint in the structured vacuum field (SVE). Just like a groove in a record or a path worn into the ground from repeated steps.

That ripple doesn't just pass—it alters the readiness of that region to respond again.

This is memory—not as neurons or storage—but as field predisposition.

A ripple zone that once aligned in a specific way is now more likely to re-align similarly in the future. This is the secret behind habit formation in systems, both physical and biological.

Gravity wells, magnetic fields, inertia = stored alignment history

Think of gravity—not as a pull, but as a compressed ripple memory well. Mass is a zone of repeated ripple compression. And the field surrounding it remembers how to collapse inward.

Magnetism? A polarity memory—a torsional ripple history attempting to resolve contradiction.

Inertia? It's not an object's resistance to change. It's the ripple path history reinforcing the same alignment behavior over time.

Even planetary orbits are not forces or trajectories. They're reinforced ripple memories—stable paths that require less field work to sustain.

You don't need strings to hold things in place. You just need memory in the field to keep re-aligning the same way.

No central processing—just persistent ripple logic

There is no master processor in the universe. No divine observer holding a clipboard.

Everything that "persists" is the result of ripple feedback memory.

When a field gets disturbed, it doesn't reset blindly. It references previous alignment states to restore balance.

That's how rules emerge—not because they were imposed, But because they were successful ripple patterns remembered and reused.

Even atoms—supposedly identical particles—aren't objects. They're ripple profiles replayed under specific field conditions.

The same hydrogen doesn't exist. The same hydrogen keeps reappearing, because the field remembers how to form it.

"Reality is what the field refuses to forget."

We think the past is gone. But in UED, the past is stored as ripple pressure in the now.

When you drop a pebble in a pond, the ripple fades. But in structured vacuum energy—it etches the field. That memory doesn't decay—it influences all future ripple behavior in its zone.

This is why consequence matters more than cause. Causes are guesses. But consequences become memory logic for the universe.

You are not held together by particles. You are held together by what your ripple field insists on repeating.

Everything you see, touch, and feel is a consequence that the field refused to forget.

6.7 Consequence > Cause: The Reversal of Causality

"What happened is not what caused it—it's what refused not to happen."

In classical science, causality is the prized compass—A makes B happen, B makes C happen. It's clean. Linear. Intuitive.

But it's also—fiction.

Because the only thing we ever truly observe is consequence. The rest—causes, reasons, origins—are stories we reverse-engineer to make sense of it all.

UED flips the arrow of causality. It declares: Nothing happens because of a cause. It happens because of the conditions that made that consequence the only viable ripple outcome.

Cause is always a retro-label on observed consequence

We never see "cause." We see an event. Then we go backward in our reasoning to label something as its cause.

If a tree falls, we say: "The wind blew too hard." But UED would ask: What field misalignments accumulated to make the collapse inevitable?

The branch didn't fall because of one gust. It fell because its ripple integrity could no longer realign with the field pressure it experienced. The fall was a stored consequence, not a singular cause.

Causality is often nothing more than our mental shorthand for ripple sequences we didn't track in full.

UED: Only consequence is measurable, repeatable, real

In a ripple-based universe, what matters is not how it started—but how it unfolded. You cannot measure a cause. You can only measure a change. And if that change repeats in predictable ripple conditions, we call it a law.

But it's not a cause. It's a ripple outcome pattern—what the field does when it remembers how to react.

This is why UED doesn't chase "first causes." It studies ripple dynamics: how fields respond, store, and shift based on prior alignments and equilibrium breaches.

Causality is not linear—it's field resonance

Let go of straight lines. Nothing goes A → B → C in the universe. Ripple fields resonate—they echo, overlap, feedback, and interfere.

Sometimes, a consequence emerges long after the "cause" is forgotten—because the field tension stored it like memory debt. Sometimes, multiple events happen at once—because the same ripple alignment permits their simultaneous emergence. And sometimes, things happen with no clear prior trigger—because the field was primed for it.

UED replaces causality with resonance logic: Consequences align when the field permits, not when logic dictates.

"What happened is not what caused it—it's what refused not to happen."

This is the final blow to mechanistic thinking.

You are not caused. You are consequential—a ripple result that persisted through memory, alignment, and delay.

Gravity didn't pull you. Ripples refused to stay unaligned.

The light didn't travel to you. The ripple became observable as the field allowed it to appear in sequence.

History is not a line of causes—it is a pattern of ripple permissions.

And you don't exist because something caused you. You exist because your ripple alignment refused not to happen.

That's the only truth UED defends: Reality is not made by triggers—it is shaped by what consequences the field allows to survive.

6.8 Closing: Why UED Makes Everything Simpler, Not Just Deeper

"There was never a thing. Only its consequence."

We began with particles. Then we added forces. Then fields. Then spacetime. Then paradoxes, dualities, contradictions, and quantum weirdness— Each new layer meant to fix what the last one couldn't explain.

Universal Energy Dynamics (UED) says: Stop. You don't need layers. You need clarity.

Because beneath every contradiction was one simple truth we ignored:

The universe is not made of things. It is made of ripple consequences.

One Field. Many Outcomes.

At the heart of UED is a stunning minimalism:

You don't need multiple forces. You don't need four fundamental interactions. You don't need matter and antimatter, particles and waves, observers and systems.

You just need one structured field—SVE—and a logic of ripple:

- Ripple emission creates form
- Ripple alignment sustains it
- Ripple delay creates time
- Ripple memory stores interaction
- Ripple collapse resets equilibrium

Everything else—mass, gravity, magnetism, charge, even thought—are emergent behaviors of this ripple choreography.

No More Dualities: Wave vs Particle, Space vs Time, Force vs Energy

Physics has spent a century trapped in dualisms:

- Is light a wave or a particle?
- Is gravity a force or a curve?
- Is time flowing, or is it blocky?
- Are electrons real, or are they clouds of probability?

These questions are symptoms of a flawed framework.

UFD deletes the dualities. Why? Because there is no thing to split.

- What we thought were waves and particles are just different ripple alignment outcomes.
- What we called space and time are just field coordinates and delay patterns.
- What we call force is just ripple pressure realigning to relieve memory tension.

You don't need opposing models when the truth is behavioral, not structural.

Just Ripple Choreography in Structured Vacuum Energy

The entire cosmos is a dance in silence. No object is moving. No force is pushing. No time is ticking.

- Only ripples—emerging, aligning, delaying, collapsing.
- Only SVE—holding potential tension, waiting to be disturbed.
- Only you—as an observer, experiencing the patterns as reality.

This is not mystical. It's not poetic license. It's field logic.

"There was never a thing. Only its consequence."

This is the final insight of UED:

There was never a particle—just a node of persistent ripple behavior. There was never motion—just a handover of alignment across field shells. There was never time—just a gap between field response and recognition. There was never force—just a compensation in ripple tension.

Even *you* are not a thing. You are a complex ripple formation with memory—temporarily stable, beautifully emergent, and consequentially real.

UED does not make the universe more complex.

It makes it inevitably simple.

Because when you stop asking, "What is this thing made of?" And instead ask, "What consequence keeps happening?"

You've stepped out of illusion—and into the field.

What Now Stands in Place

We spent five chapters disassembling the scaffolding of classical and modern physics. What was left? Not void—but vibration. Not particles—but pressure. Not cause—but consequence.

Here's what now stands in place:

- **Particles?** Merely ripple alignments—zones where memory stabilizes into repeatable form.
- **Motion?** Just ripple handover—no object ever moved, only alignment shifted.
- **Force?** A mirage—what we felt was ripple resistance to misalignment.
- **Time?** Not flowing—just the illusion created by ripple delay and memory lag.
- **Spacetime?** Not a fabric—just compressed ripple zones adapting to field pressure.

What UED offers is not a patch, but a rewrite of the canvas: Reality is not built from parts. It is composed of field consequences that align, delay, remember, and balance. Everything you see, sense, and measure—is not a cause. It's a replay of imbalance searching for resolution.

You are not made of matter. You are a ripple performance stabilized for a moment. There is no motion—only refresh. There is no force—only restoration. There is no time—only the rhythm of re-alignment.

And at the base of it all? Structured Vacuum Energy—the poised silence from which the song of reality begins.

This chapter didn't introduce a theory. It introduced reality—as it always was, before names, particles, or patches.

Chapter 7: Light Doesn't Travel

Why the Sun Never Sends You Anything

Let's dismantle the foundational illusion that light moves through space. This chapter shows that what we call "light" is not a traveling particle or wave—but a ripple alignment event that triggers perceptual consequence. UED reframes light as a ripple handshake between sender field and observer field—no photon ever left the source.

They told you light travels. That the Sun sends out endless streams of photons shooting across the vacuum, bouncing off planets, sliding into your eyes. You've heard it so often, it feels like truth carved into the cosmos.

But what if the Sun never sent you anything?

What if what you call *light* is not a message in flight—but an agreement, a shared alignment across ripple fields? What if your eyes never *received* anything, but instead *resonated* with a field that had already reached readiness?

In Universal Energy Dynamics, light doesn't move—it manifests. Not as a bullet. Not as a wave. But as a consequence of fields aligning across distance—without anything *traveling* at all.

This chapter will illuminate the darkness in our understanding. Not with photons—but with alignment.

Because you didn't catch the light. You completed the handshake.

7.1 The Great Misunderstanding

"We never caught light moving. We just saw consequence appear."

For centuries, light has been misunderstood—not because it was elusive, but because we were obsessed with explaining it in terms we already understood: objects, forces, trajectories. The mind craves movement, direction, and cause. And so, from the earliest theories to today's textbooks, light was forced into roles it never auditioned for.

In classical physics, light began as rays—imaginary straight lines cast by luminous bodies. This model served navigation and shadows well enough but offered no substance. Then came Newton's corpuscles, tiny particles of light shot from sources to observers like invisible bullets. A neat analogy, but it failed when faced with phenomena like diffraction or interference.

To patch this, Huygens proposed light as a wave, spreading through an imagined "aether," an invisible jelly no one could detect. That model worked for interference—but failed for photoelectric effects. Enter Einstein's photon, reviving the particle but attaching energy to frequency—a hybrid solution.

This back-and-forth became the wave-particle duality, not as a truth, but as a desperate compromise between models that each broke down under different tests.

But at no point did we ever observe light *traveling*. What we always measured was the *outcome*—a change, a response, a signal arriving.

We saw leaves warm. Screens glow. Eyes react. And from these consequences, we imagined a cause—a traveling something.

But what if that cause was never real?

What if "light" is not the actor, but the aftermath?

Not a thing en route, but a ripple agreement between distant fields?

What if the only reason we think light moves...is because *we never questioned why it appeared*?

7.2 No Body Emits Light

"Emission is ripple provocation. Perception is ripple resolution."

We've long imagined the Sun as a grand emitter—hurling photons across the vacuum like fireballs of energy. In classrooms and textbooks, we're taught that atoms get "excited," then "emit" light when they return to ground state. The vocabulary is theatrical: emission, release, radiation, transmission.

But this picture, though visually satisfying, is fundamentally flawed.

No physical object emits anything across space. Not the Sun. Not atoms. Not bulbs. Not your screen.

Instead, what we call "emission" is merely a disturbance in the structured vacuum field (SVE)—a ripple provocation that registers as alignment only when another ripple field (like your eye or a sensor) is in receptive phase.

Think of it this way: you don't receive a message because someone "threw" it across space—you receive it because your field resonated with the disturbance that happened elsewhere.

The Sun doesn't launch photons. It provokes local ripple turbulence through nuclear reactions—those ripples never leave in the classical sense. They spread tension through the SVE, just like tapping one end of a tight fabric sends pressure to the other—but nothing ever detaches and travels.

When you "see" the Sun, you are not receiving particles. You are experiencing ripple agreement—a mutual handshake between the ripple condition at your local field and the memory encoded in the universal field.

And when atoms "emit," they don't send out little marbles of light. They simply enter a new ripple pattern, one that creates perceivable consequence if an observer's ripple field is in compatible phase.

In Universal Energy Dynamics, there are no projectiles. There are ripple provocations, field alignments, and consequence triggers.

Light is not emitted. It emerges—wherever and whenever the conditions agree to resolve tension.

7.3 The Observer's Role: Why Light Depends on You

"You didn't catch the photon. You completed the handshake."

We've spent centuries believing light travels independently from source to destination—as if it exists "out there," waiting for us to catch it. From Newton's rays to Einstein's photons, the observer has been portrayed as a passive receiver of incoming information.

But in the UED framework, that entire paradigm collapses.

There is no light unless you complete it.

What you call "light" isn't a substance moving toward you. It is a ripple consequence, triggered only when your own field aligns with the disturbance from another region of the Structured Vacuum Energy (SVE). In this sense, vision is not passive detection—it's active ripple agreement.

Let's be clear: No photon exists flying through space. There is no particle in transit. What exists is a patterned pressure field, expanding from the source—waiting to resolve when a local field (like your eye) becomes resonant.

Your eyes don't "receive light"—they collapse alignment paths into perceivable experience. Your retina doesn't catch anything. It responds to ripple choreography by finalizing a mutual condition. This is not reception; it is co-creation.

You don't "see" because something traveled. You see because something was completed.

This is why blind spots exist. Why different observers perceive differently. Why certain wavelengths are "invisible." Because unless your field is attuned, no amount of ripple pressure will trigger visibility.

Just like tuning a radio doesn't pull a signal out of space—it locks into an existing field structure—your act of perception is an act of ripple alignment. Nothing is seen unless you are structurally aligned to resolve that tension.

In UED, there is no light without an observer. Not in a quantum sense of "collapse"—but in a field logic sense: The ripple consequence simply doesn't actualize unless a receptive field completes the cycle.

So no—You didn't catch the photon. You were part of the contract that made light visible in the first place.

7.4 Alignment Threshold Time (ATT)

"Light didn't arrive in 8 minutes. It required 8 minutes of matching conditions."

One of the most cherished constants in modern physics is the speed of light (c)—an untouchable value of ~299,792 km/s, considered not just a speed limit, but the foundational yardstick of reality. It defines how fast light "travels," how time dilates, and how spacetime bends.

But UED doesn't need light to travel at all.

Instead, it introduces something far more grounded—and far more revealing: ATT—Alignment Threshold Time.

What is ATT?

ATT is not a velocity. It's not a measure of how fast something moves through space.

It is the minimum time required for ripple fields to establish coherent alignment between the source and the observer. In other words, ATT is the ripple synchronization delay—the time it takes for conditions to become precisely aligned enough for a consequence (like the perception of light) to emerge.

So when we say it takes 8 minutes for sunlight to "reach" Earth, UED reframes it:

The field conditions between the Sun and Earth took 8 minutes to reach the alignment threshold required for that specific ripple to be resolved as light in your field.

No photon raced through a vacuum. No wave oscillated across space.

Instead, there was a continuous field of ripple pressure between Sun and Earth. It simply took 8 minutes of propagation delay—like passing a wave through a long rope—to fully align the field nodes from emission to perception.

What Sets This Delay?

ATT depends on:

- Ripple density of the vacuum field
- Distance between nodes of alignment
- Degree of distortion or resistance in the intermediate field
- Observer's field readiness to complete the handshake

This delay is universal, not because it defines a speed—but because it defines a baseline ripple responsiveness across SVE. The so-called speed of light is thus an emergent constant, arising from the minimum threshold delay needed to create a visible ripple response.

In simple terms:

Light speed is ripple patience—not ripple velocity.

Rethinking the Constant c

Under UED, c is not a fundamental speed limit. It is the structural minimum ATT for the field to generate perceptual coherence. It tells us how long it takes for a ripple agreement to form between two stable zones under ideal vacuum conditions.

This also explains:

- Why nothing outruns light: because ripple coherence must reach its threshold before anything appears.
- Why light seems delayed but never observed en route:

because the ripple doesn't show itself until the handshake is complete.

- Why photons are never seen mid-flight: because they don't fly. They emerge at resolution points.

So when you gaze at a star millions of light-years away...

You're not seeing something that traveled. You're seeing a field finally reach resolution—after waiting eons for alignment.

That is the elegance of ATT: Reality appears not when something arrives, but when something aligns.

7.5 Why There Is No Medium, No Wave, No Particle

"We don't need a carrier. We need agreement."

For centuries, science has chased "the thing that carries light." Was it a medium? Was it a particle? Was it a wave?

Each century installed a new scaffold to uphold the illusion:

The Ether: A Medium Without Proof

Early physicists imagined ether—a ghostly substance permeating space, carrying light the way air carries sound. It made intuitive sense: waves need something to ripple through, right?

But experiments like Michelson–Morley failed to detect it. So ether was declared dead, and light was rebranded as self-sufficient: a wave *without* a medium.

The Photon: A Patch on a Patch

To explain light's interaction with matter (like photoelectric effects), Einstein proposed photons—tiny, indivisible light packets. These "particles of light" carried energy and momentum.

But wait—wasn't light also a wave? Enter:

Wave-particle duality—an uneasy marriage between contradiction and convenience.

Now light could behave like a wave when diffracting, a particle when detected, neither when questioned, and both when misunderstood.

The photon didn't solve the mystery—it decorated it with paradox.

The Wave: Misinterpreting Interference

Light's wave behavior was treated as proof of "oscillation"—but oscillation of what?

- There's no known medium.
- Fields don't have volume to oscillate.
- A photon is too small to wave.
- And nothing was ever observed *in transit*—only at origin and detection.

Wave interference, in reality, is ripple choreography—not undulating substance.

We never proved a wave; we witnessed patterns and reverse-engineered the wrong narrative.

Enter UED: No Carriers. No Contradictions.

Universal Energy Dynamics removes the need for:

- A medium (ether)
- A carrier (photon)
- A structureless wave (undefined oscillation)

Instead, it posits a Structured Vacuum Energy (SVE)—a stable ripple field that, when disturbed, generates ripple sequences. These sequences propagate not as particles or waves but as field alignment pressures.

Nothing travels. The field reconfigures. Nothing carries. The field agrees.

Agreement Over Movement

When light is perceived, it's not because a particle flew into your retina. It's because your field finally reached coherence with a ripple emitted from a remote source.

Perception is field agreement—not delivery.

There is no photon zooming through vacuum. No "bullet of energy." No photon box.

So we drop the wave. We drop the particle. We drop the need for medium.

And what's left?

A reality framework that's ripple-aligned, self-consistent, and free of duality.

UED doesn't patch the paradox. It dissolves the need for one.

7.6 What Vision Really Is

"You don't see with eyes. You align with the ripple echo."

For all the brilliance attributed to the human eye, we've misunderstood what it's actually doing. We treat vision like photography—as if the eye captures an image, passes it to the brain, and voilà—"seeing" happens.

But in UED, vision is not capture. It's not detection. It's resonance.

The Retina Doesn't Capture—It Tunes

Your retina is not a screen. It's a living ripple interface—an organic array of sensors tuned to align with ripple patterns in the surrounding Structured Vacuum Energy (SVE).

- Rods and cones don't receive photons.
- They resonate with incoming ripple shifts.
- These are not particles striking tissue—they're pressure alignments triggering a biochemical handshake.

You don't receive light. You agree with it.

The Brain Doesn't Process Images—It Resolves Alignment

The visual cortex doesn't translate "data from the eye." It resolves ripple echo agreements into recognizable consequence. What we call "image" is a high-probability reconstruction of ripple resonance.

This isn't passive viewing—it's active ripple negotiation. You're not staring at a tree.

You're entering field compliance with the tree's ripple memory.

No Light Enters the Eye

Let this sink in:

No photon enters your body. What enters is a synchronized ripple pattern, matched by a biological structure that has evolved to recognize, stabilize, and interpret that pattern.

You're not "seeing the Sun." You're feeling a consequence field that your eye-brain system has learned to co-resonate with.

Sight Is Not Input—It's Conformity

Vision is not the reception of energy.

It is:

- The alignment of ripple signatures,
- The resolution of pressure coherences,
- And the interpretation of echo consequences by a structure that's been biologically calibrated to do just that.

You don't see objects. You perceive field continuity through ripple settlement.

You see only what your ripple structure allows.

Summary Insight

- Vision is not based on what's "out there."
- It's entirely dependent on what aligns in here (within your biological ripple frame).
- Every "image" is a ripple agreement—not an object's broadcast.

So stop asking "what hit the retina?"

Instead, ask: What field conditions aligned such that perception emerged? Because in UED, perception is not a reaction. It's a coincidence of structure and ripple echo.

7.7 Experiments Reinterpreted

"Every experiment saw behavior, not entity. Every result was consequence, not proof."

Physics became obsessed with proving what "light is." But what if we were never observing light at all?

What if every experiment was merely recording ripple consequences under different boundary alignments?

In Universal Energy Dynamics (UED), no experiment ever captures a thing. It reveals how the ripple field behaves when disturbed, constrained, or measured.

Let's revisit the famous experiments—not as confirmation of particles or waves—but as rituals that provoke ripple responses.

Young's Double Slit: Not Interference—Alignment Ambiguity

Mainstream View:

- Light behaves like a wave—it interferes.
- But when observed, it "collapses" into a particle.
- Welcome to duality confusion.

UED View:

- Ripple fields align through both slits simultaneously.
- The pattern on the screen isn't interference—it's probability density of ripple agreement.
- When we try to observe which slit "the photon" passed through, we change the field pressure, collapsing the ripple alignment prematurely.

There's no photon to pass through either slit. There's only field tension finding its best match on the detection plane.

No wave. No particle. Just ripple logic.

Photoelectric Effect: Not Particle Proof—Ripple Threshold

Mainstream View:

- Only photons above a certain energy knock electrons out.
- Therefore, light is a packet (photon).

UED View:

- The surface has a ripple resistance threshold.
- A ripple of sufficient compression coherence activates electron release—not because it's a particle, but because it overcame field stability.

Frequency works not because of "more energy per photon," but because higher-frequency ripple disturbances have sharper compression signatures—which breach the field stability of the material.

No photon hit. Just ripple handshake—field to field.

Polarization: Not Direction of Oscillation—Phase Agreement Constraint

Mainstream View:

- Light "waves" have an orientation. Polarization filters block or pass based on this.

UED View:

- Ripple emissions have phase orientation due to source geometry.
- Filters are not blocking "wave direction"—they're

modifying the phase-match conditions required for the ripple to continue alignment into the detection zone.

Polarization is not wave directionality— It is field phase agreement or rejection.

What All These Prove

None of these experiments prove particles. None confirm waves. They all demonstrate:

- Ripple emergence
- Alignment thresholds
- Field memory consequences
- Observer-dependent agreements

In short: behavior, not being.

Final Insight:

A behavior is not a proof of what exists. It's a reflection of what the ripple field allowed under constraint.

So every Nobel-winning experiment? It didn't prove the nature of light.

It revealed how ripple fields behave when:

- Disturbed,
- Aligned, or
- Observed out of equilibrium.

7.8 Closing: The Sun Never Left

"You didn't see the Sun. You felt the agreement."

Let's pause and remember what we've just undone.

For centuries, we believed that light *traveled* from the Sun to Earth. That it raced across space as particles (photons) or waves—*emitted, sent, received.* But not once did we question the assumption beneath the metaphor: What if the Sun never really *sent* anything?

The Sun Did Not Emit—It Aligned.

In UED, the Sun is not a transmitter. It is a ripple node—a massive, compressed alignment center in the Structured Vacuum Energy (SVE) field.

Its presence disturbs the SVE field persistently, setting up ripple tension gradients throughout surrounding space.

Your body, your eyes, your instruments—they do not receive sunlight.

They complete the ripple agreement required for perception. They resonate with a ripple field that was already there.

No "photon" left the Sun. No "wave" traveled for 8 minutes. What reached you was the readiness to align—when your field finally synced with the Sun's ripple memory.

8 Minutes Was Never a Journey

We say, "Sunlight takes 8 minutes to reach us." That's not a delay in movement—that's the minimum time required for ripple conditions across space to sync and allow the illusion of light to emerge.

What actually took 8 minutes?

- Not a particle's travel.
- Not a beam in motion.

It took 8 minutes for ripple alignment across trillions of microscopic phase conditions to echo in your eye.

It wasn't sent. It was permitted—field to field, ripple to ripple.

The Big Shift

Light isn't:

- Emitted
- Propagated

- Absorbed

It is:

- Ripple tension in memory fields
- Consequential resonance across structured vacuum
- Perception born from alignment

One Last Time:

You did not *see* the Sun. You did not *receive* its rays. You felt the agreement—when your ripple field was ready to echo its consequence.

The Sun never left. You just aligned with its presence.

You Never Saw Light—Only Its Consequence

A Final Reflection

In this chapter, we shattered the most cherished illusion in physics—that light travels.

Here's what we left behind, and what we replaced it with:

We shattered the belief that...

- Light is a *thing* that moves from one place to another.
- The Sun emits photons that *travel* across vacuum.
- Vision is a passive act of *receiving particles.*
- Light speed is the *velocity of an entity.*
- Experiments prove *wave-particle duality.*

We revealed instead that...

- Light is not a traveler—it's a ripple consequence that only appears when the observer's field is ready.
- The Sun does not send anything—your field simply echoes its ripple imprint.
- What you call "seeing" is alignment—a field-to-field handshake, not a beam-to-eye delivery.
- "Light speed" is Alignment Threshold Time (ATT)—not motion, but minimum synchronization delay.
- Classical experiments measured effects, not entities. They recorded how fields realign, not what travels.

The New Understanding:

- No photon exists independently—only ripple disturbance and alignment do.
- There is no medium, no wave, no particle—only Structured

Vacuum Energy rebalancing.

- Vision is ripple memory resolution. You don't *capture light*—you complete its pattern.
- The Sun never moved toward you. You never caught it. You simply aligned.

What Light Truly Is:

"A ripple alignment in the universal field, echoed by your existence, felt as a moment of presence."

This wasn't just about optics. It was about letting go of an entire worldview. You never saw light.

You saw consequence. And that changes everything.

Chapter 8: No Source, No Field, No Force

Why Nothing Pushes—It All Remembers

We've been chasing phantoms. For centuries, we believed that objects emitted forces—that masses pulled, charges pushed, and magnets reached out across space to grab or repel. We drew invisible lines of influence called "fields" and anchored them to "sources," like celestial bureaucrats issuing orders across the cosmos.

But what if none of that was ever real?

In Universal Energy Dynamics (UED), there is no source projecting anything. There is no force acting upon another. There are only ripple tensions within Structured Vacuum Energy—stretching, compressing, or relaxing—like a cosmic memory foam redistributing pressure.

What we interpreted as magnetism, gravity, or radiation are not "forces" beaming out of objects, but consequences of the field's own attempt to return to equilibrium. They are the echoes of ripple contradictions, not projections from matter.

This chapter dismantles the very foundation of classical and modern field theory. It reframes everything we thought came *from* something as merely realignments *within* something deeper.

There is no push. There is no pull. There is only remembrance.

Let's step into a universe that never acted—but always responded.

8.1 The Source Illusion

For centuries, physics has worshipped the "source."

We've been taught that every field must originate from something:

- A magnet emits a magnetic field.
- A charged particle emits an electric field.
- A planet or star emits gravity.

This narrative has shaped every branch of physical theory. It gave us point charges, mass centers, field vectors, and source equations. But here's the disruption: *There is no source. There is only structure.*

What we call a "source" is merely the most visible point of imbalance in a ripple configuration. It's a marker, not a generator. The ripple field—Structured Vacuum Energy (SVE)—is already everywhere, and the so-called "source" is just a region where ripple tension peaks, where field memory is densest, or where alignment has been disrupted.

In UED, the field isn't emitted from an object. The field is the object—extended, stretched, and stabilized into what you perceive as influence.

That magnet on your fridge? It's not "producing" anything. It's a frozen ripple contradiction, causing surrounding space to reorganize in tension. That charged particle? It's just a ripple misalignment, not a central engine of force.

The illusion of a source arose because we wanted a start point—something to blame or glorify. But ripple logic doesn't require a central authority. It requires a memory. A configuration. A geometry of consequence.

So let's stop asking "where does the force come from?" The better question is: "What ripple contradiction is the field resolving right now?"

"You called it a source because you didn't understand the field's memory."

8.2 There Are No Fields—Only Conditions

What we traditionally call a field—be it magnetic, gravitational, or electric—is not a *thing* floating in space. It is not an invisible fabric that wraps around objects and acts upon others.

In UED, there is no such *substance* as a field. What we observe—and mislabel as a "field"—is actually a zone of ripple imbalance, a conditional geometry of Structured Vacuum Energy (SVE) undergoing tension, compression, or delay.

Let's break the illusion:

- A magnetic field is not something a magnet emits. It is the local consequence of ripple phase contradiction, trying to resolve opposite alignments.
- A gravitational field is not a pull extending from mass. It's a compression basin in ripple memory—a region where surrounding alignments collapse inward to relieve built-up strain.
- An electric field is not radiating energy. It's the sharp slope of ripple misalignment, a region where ripple shells are under forced reconciliation.

All these are just conditions—ripple responses to imbalances in the SVE continuum. They don't have independent identity. They're not "forces out there." They are consequences appearing here.

And most importantly: They are not projected—they are provoked.

Whenever we say "this region has a strong field," we really mean: "This region is under high ripple memory tension. Alignment is more expensive here."

So forget "field lines" and "force carriers." That was visual poetry, not ontological reality.

What you're witnessing is organized imbalance behavior. A field is not something added to space. It is space trying to forget what disturbed it.

"A field is not a thing—it's the behavior of imbalance."

8.3 Magnetism as Phase Tension

We've long believed that magnets "pull" or "repel" objects—casting invisible force lines into the void. But UED strips away this illusion by revealing the deeper mechanism: Magnetism is not a force. It is ripple phase tension.

Let's decode the illusion:

- Poles are not ends of a magnetic object emitting mysterious power. They are zones where the ripple phase of the object is misaligned with its surrounding Structured Vacuum Energy (SVE).
- A north pole is not pushing or pulling. It's a region of ripple phase attempting to converge with another. The south pole represents the complementary phase. When opposite poles meet, their ripple phases find synchrony—what we call attraction.
- When like poles approach, their ripple phase alignment conflicts, causing a tension the field resists—so the system seeks to avoid overlap, which we perceive as repulsion.

The Truth Behind the Phenomenon

There is no energy emitted. No magnetic "lines" wrapping around a bar or planet. No "field" being thrown outward.

Instead, the SVE surrounding a magnet is in a state of ripple phase distortion, holding onto a memory of imbalance. Every time another object with ripple structure enters this zone, the two ripple sets try to negotiate alignment.

- If convergence is possible, we see attraction.
- If phase contradiction intensifies, we see repulsion.
- If the field is already aligned, we see neutrality.

This is not action-at-a-distance. This is co-presence of ripple memory tension, resolving itself locally through synchronized behavior.

So the magnetic force is not pushing or pulling.

It is SVE attempting to cancel its own contradictions by adjusting ripple alignment.

What we measure as magnetic strength is the intensity of unresolved ripple phase contradiction across that region.

"Magnets don't pull—they resolve their ripple contradiction."

8.4 Gravity as Memory Compression

Modern physics teaches us that gravity is either a pulling force (Newton) or a bending of spacetime (Einstein). But what if nothing pulls? What if nothing bends?

What if gravity is simply what happens when ripple memory becomes compressed—and everything nearby surrenders to that tension?

The UED Perspective

In Universal Energy Dynamics (UED), mass is not a thing with pull—It is a ripple node, where Structured Vacuum Energy (SVE) is densely aligned, tightly packed, and compressed with memory.

This compression alters the local ripple conditions—creating a gradient of tension in the surrounding field.

Now, nearby ripple systems (like your body, a falling apple, or even photons) respond not because they are pulled, but because they experience less resistance by aligning into the compressed memory gradient.

Gravity Is Field Consequence

- Gravity is not a force. It is a reaction—the field trying to resolve uneven ripple alignment around a compressed node (what we call mass).
- The so-called "fall" toward Earth is not descent in space—It is a realignment through ripple memory gradients, where field tension is lower.
- This explains why gravity acts "downward" on Earth: The center of mass compresses ripple memory most intensely there, so everything aligns inward, toward minimum ripple resistance.

No Pull. Just Preference.

- A leaf falling, a comet spiraling, or your body resting on a chair—All are outcomes of ripple choreography, not of invisible strings.
- The illusion of force exists because we misunderstand ripple alignment as movement or pull.
- Just like iron filings align to a magnet—not because they're pulled, but because the field demands less resistance in alignment—similarly, objects surrender into gravitational fields because that's the easiest path of ripple compliance.

Key Insight: Gravity is not a law. It's a memory field organizing itself, and you are part of that reorganization.

You are not "being pulled" by Earth. You are finding rest in Earth's ripple compression zone.

"You don't fall toward Earth. You surrender into its ripple compression."

8.5 Radiation as Field Relaxation

Why the Universe Doesn't Throw—It Exhales

UED Reframe: Radiation Is Letting Go, Not Sending Out

Classical physics teaches us that radiation is energy projected outward—from a star, a heater, or a charged body. But in Universal Energy Dynamics (UED), nothing is projected. Nothing is fired, flung, or sent.

Instead, radiation is the natural process of ripple field relaxation—a release of misalignment, a settling down of ripple tension into Structured Vacuum Energy (SVE) equilibrium.

Emission ≠ Projection

- There is no push, no ejection. Radiation is not a projectile event—it is a ripple rebalancing act.
- When a body is excited—say, heated or energized— its local field stores tension in ripple misalignment.
- Eventually, that misalignment becomes unsustainable, and the field begins to relax by releasing excess ripple interference.
- This release is what we call light, heat, or electromagnetic radiation—but it is not energy leaving, it is energy resolving.

Heat, Light, and Charge as Relaxation Paths

- Heat is not vibration—it's ripple agitation trying to return to coherence.
- Light is not emitted photons—it's field-level handshake consequence between unstable and stable ripple nodes.
- Charge leakage is not loss—it's ripple discharge, a return toward equilibrium in the surrounding SVE.

Radiation, in all forms, is not a forceful act. It's a surrender—like a sigh after strain.

No Carrier Needed

Radiation doesn't need a medium or a particle to carry it. It simply unfolds through ripple alignment across field zones.

There's no light wave traveling, just zones relaxing in sequence—like ripples spreading across a memory-loaded lake.

Key Consequences

- Radiation is passive release, not active emission.
- It only occurs when the ripple strain exceeds containment threshold.
- It propagates only when alignment pathways are available in the surrounding field.

"Radiation is not thrown—it's let go."

The universe doesn't explode energy outward. It simply relieves ripple pressure—gently, quietly, and in perfect alignment.

8.6 Why No Force Ever Existed

Force Was the Name We Gave to What We Never Understood

The End of Force: A Fiction of Convenience

From Newton to Einstein to modern quantum theories, force has always been the backbone of scientific thinking. But in Universal Energy Dynamics (UED), force is not a real entity—it is merely a label we applied to unmapped ripple consequences.

We saw objects accelerate, twist, bounce, or resist—and we said something acted on them.

But nothing ever acted. Nothing ever pushed or pulled.

What Actually Happened: Ripple Memory Seeking Resolution

- Every observed change in nature—from a falling apple to a

magnetic pull—was not caused by a force.

- It was a ripple consequence, unfolding across Structured Vacuum Energy (SVE) under misalignment tension.

In short: No object caused another to move. Instead, each field zone responded to stored imbalance—a ripple memory trying to resolve itself.

Action as Reconciliation, Not Impact

- What classical physics calls "action" is actually realignment. A movement toward relief from tension, not exertion of power.
- A stretched spring doesn't "push"—it returns to alignment.
- A falling object doesn't "get pulled"—it descends into lower ripple compression.
- A spark doesn't "jump"—it releases ripple strain into the nearest compatible zone.

The Real Equation: Memory > Mechanism

There was never a mechanism of force—only the ripples of misalignment expressing themselves.

What we called F = ma was just an attempt to predict behavior. But it never explained why that behavior occurred. UED does.

From Force to Field Tension Maps

- Force is not fundamental—field tension is.
- What we once called "external force" is actually internal compensation across a unified field.
- Reality isn't built on impact—it is built on field reconciliation.

"Nothing acted on anything. Everything aligned toward relief."

That's the truth behind every motion, every shock, every spin. The universe never used force. It only remembered imbalance—and moved to correct it.

8.7 Closing: What Was Never There

There Was No Source. No Field. No Force.

The Grand Disappearance of Power

After centuries of dissecting the universe through lenses of force, field, and source, we stand at the edge of a paradigm shift.

And here's what we now see: There was nothing there.

No force ever pushed. No field ever emerged from an object. No source ever emitted a thing.

These were mental scaffolds—tools of a fragmented understanding. Concepts born from misinterpreting ripple consequences in a vast, structured equilibrium we never stopped to question.

What Actually Exists: Ripple Logic in Structured Vacuum Energy (SVE)

- The Structured Vacuum Energy wasn't a passive void—it was an active framework of balance.
- Everything we called energy, radiation, or interaction was nothing but the resolution of ripple tension.

Reality was not built by throwing forces around—it was maintained through silent, synchronized realignment.

No Action—Only Accounting

- What we labeled as magnetism, gravity, or radiation was simply field memory working to restore ripple equilibrium.
- The universe is not assertive. It is accountable.
- No entity commands the cosmos—it all unfolds because the field refuses not to balance.

From Power to Peace

If classical physics was about who acted on whom, UED is about how everything seeks rest.

"The universe doesn't exert power. It restores peace."

This isn't metaphor. It's mechanics. It's not mysticism. It's structured logic.

The cosmos does not run on power dynamics. It runs on ripple debt and its resolution—no more, no less.

And So, What Was Never There?

No force. No push. No pull. No source.

Only ripple consequence. Only memory realigning. Only a universe trying to be whole again.

What We Let Go Of

You Were Never Pulled—You Aligned

We've spent generations believing we lived in a universe full of pushing, pulling, throwing, and charging. We invented invisible fields, pointed to magical sources, and labeled anything unresolved as a force.

But now we let go.

- Let go of "force"—it was only the discomfort of imbalance.
- Let go of "source"—nothing ever began at a point.
- Let go of "field"—it wasn't a thing, just a ripple behavior.
- Let go of action—there was only memory trying to settle.
- Let go of power—the universe never needed it.

You were never pushed. You were never pulled. You were never acted upon. You aligned. You synchronized. You relieved tension in a ripple field trying to become whole again.

UED doesn't ask you to believe in a new story. It simply clears away the old noise, so you can finally hear the hum beneath it all— A universe that remembers instead of reacts, Realigns instead of resists, And restores peace instead of imposing power.

You were never pulled. You just returned to where you belonged.

Chapter 9: Day, Night, Seasons, and the Still Earth

Why the Sky Changes—Not the Earth

We were taught to believe that Earth is spinning at a thousand miles per hour, hurtling around the Sun at 30 kilometers per second, all while tilting and wobbling its way through the cosmos. And yet—we never feel a thing.

Why is that?

Because the Earth isn't moving.

Not in the way we've been told. Not in the way Newton or Copernicus imagined. Not even in the relativistic daydream Einstein spun. In the lens of Universal Energy Dynamics (UED), Earth is a still ripple node—a stable alignment knot embedded within Structured Vacuum Energy (SVE). It is the field that breathes, not the rock beneath your feet.

Day and night don't happen because Earth rotates—they happen because ripple alignment conditions shift across the surface field. Seasons don't come from orbital ellipses—they arise from cyclical tension memory in the ripple lattice, slowly recycling alignment harmonics.

The Sun never moves across the sky. Your perception of the ripple agreement moves—and you call it sunrise.

The sky changes because you align differently. You don't spin, orbit, or chase the Sun. You stay still—and everything else ripples.

This chapter unhooks you from the planetary merry-go-round and restores your footing—not just on Earth, but in reality. Let's begin.

9.1 The Rotating Earth Assumption

For centuries, Earth was believed to be the immovable center of the universe. Then came the Copernican revolution, followed by Newton and Galileo, who insisted that Earth spins on its axis and orbits the Sun in a celestial ballet of gravitational mechanics. This heliocentric model gained dominance not because it was proven by motion, but because it offered predictive ease for seasons, eclipses, and planetary paths.

The assumption was simple: if we experience day and night, and if the stars seem to wheel across the sky, it must be because Earth is spinning. If seasons change, Earth must be tilting and revolving. The logic followed the path of visual correlation—not foundational understanding.

But UED challenges this correlation head-on.

According to Universal Energy Dynamics, motion is not the root of experience—alignment is. Earth is not spinning or orbiting anything. It is a ripple node in Structured Vacuum Energy (SVE)—a point of stable alignment where ripple consequences converge, persist, and manifest.

You don't witness the sky changing because Earth moves. You witness it because your ripple alignment with the field changes. What we have called "rotation" is merely the shifting projection of field behavior over a stationary ripple base—the Earth.

We assumed Earth rotates because we perceived change. But we never asked—change in what? Not in position. Not in momentum. In field condition. And what if that change wasn't due to movement at all?

UED reframes the debate:

- Not "what moved?" but "what realigned?"
- Not "how fast?" but "how synchronized?"

This is not a return to geocentrism. It is a transcendence beyond all celestial mechanics—toward field logic that never required or admitted real motion in the first place.

"You didn't move. Your field remembered differently."

9.2 Day and Night Are Not Rotational Effects

We have long been told that day and night are caused by Earth's rotation—one half facing the Sun while the other rests in darkness. But what if that's not what's really happening?

In Universal Energy Dynamics (UED), nothing spins. Nothing orbits. There is no physical motion of a planetary body required to explain what we call "day" and "night." Instead, UED proposes a much simpler—and deeper—truth:

Day and night are field phenomena, not mechanical events. What changes is not your position in space, but your alignment threshold with the Sun's ripple field.

Here's how:

- The Sun's ripple influence is constantly present within the Structured Vacuum Energy (SVE).
- Observers on Earth do not receive "light" from a flying photon or a moving sunbeam.
- Instead, they come into resonant agreement with a ripple pattern emitted and sustained in SVE.

When your local field alignment crosses the necessary threshold—called the Alignment Threshold Time (ATT)—your perceptual system completes the handshake. This is what we interpret as "seeing light". That moment becomes morning.

As your local field begins to desynchronize from that ripple knot, your perception fades—evening arrives. The Sun hasn't moved, nor has the Earth spun. What changed was the field configuration necessary for perception.

In this model:

- "Noon" is not when the Sun is overhead—it's when your ripple field is in peak alignment with solar consequence.
- "Night" is not darkness caused by the Earth's shadow—it's the absence of coherent ripple agreement.

The ripple knot—the coherent zone of aligned solar ripple—moves across the Earth's stable SVE node, not because of Earth's rotation, but because of how ripple memory unfolds through field tension zones. It's a traveling synchronization wave, not a rotating spotlight.

"Daylight didn't come to you. You came into ripple agreement with it."

"Night didn't fall. Perception faded as the handshake dissolved."

Thus, morning and evening are not celestial timings—they're ripple events. They occur where your consciousness aligns or detaches from the larger field narrative.

In short, you never rotated into the Sun. You realigned into vision.

9.3 Seasons Without Orbit

For centuries, we've been taught that seasons change because the Earth orbits the Sun, with its axis tilted at an angle. As Earth moves around the Sun, different parts of the planet receive varying amounts of sunlight, causing the cyclical variations in temperature and light that we call seasons.

But in Universal Energy Dynamics (UED), there is no orbital motion. There is no Earth "spinning around" a central star. Instead, seasons are the natural consequence of how ripple memory tension cycles through the Structured Vacuum Energy (SVE).

UED Reinterpretation of Axial Tilt and Revolution

The concept of axial tilt and orbital revolution suggests that Earth's tilt is what causes seasonal variation. As the Earth orbits the Sun, different hemispheres tilt toward or away from the Sun, giving rise to the seasons.

But in UED, seasons are not driven by orbital mechanics. Instead, they are a result of cyclic ripple tension patterns that naturally emerge within SVE.

- Axial tilt is not the cause of seasonal change; it is simply a point of reference within the SVE system, determining how the local ripple alignment shifts throughout the year.
- Similarly, the revolution of Earth around the Sun does not create seasons—it merely marks the observable cycles of ripple tension unfolding in alignment with the Earth's position in the field.

The Sun doesn't "heat" the Earth because it's closer or farther away. The Earth's ripple interaction with the Sun's field creates cyclical tension discharges, which we perceive as seasonal change.

Seasons as Cyclic Memory Discharges

Seasons are not orbital phenomena—they are periodic memory discharges in ripple tension zones. Each season is a release or resolution of tension that has built up in the Earth's field alignment with the Sun.

As Earth re-aligns with solar energy throughout the year, it encounters different "ripples" in the field's memory, which manifest as:

- **Spring**: A period of release as the Earth's field relaxes into more balanced alignment with the Sun, leading to growth and renewal.

- **Summer**: A peak period of ripple resolution, when the alignment of the Earth and Sun creates the most intense ripple interaction, leading to warmth and energy.
- **Autumn**: A time of shifting equilibrium, where the tension relaxes and prepares for the cooler months.
- **Winter**: The contracted memory zone, where the Earth's alignment rests in a lower-energy state, waiting for the cycle to begin again.

Each season is a repeating memory discharge, not caused by the movement of Earth, but by the cyclical recalibration of the field's ripple behavior.

Earth Stays Still—The Field's Ripple Cycles Shift

In UED, Earth does not move through space in an orbit. It is still, remaining in its stable position within SVE. The cycles of seasons arise from the shifting patterns of ripple behavior that affect the Earth's localized field. As the field around Earth shifts in alignment, we perceive these shifts as seasonal changes.

- The Earth's position in the field doesn't change relative to the Sun in an orbital sense—it shifts in terms of how it aligns with solar memory.
- **Seasons are not a result of Earth's orbit—they are consequences of periodic ripple re-alignment within the field, which we experience as seasonal shifts.

In essence, Earth is not traveling through space, it is simply remaining aligned in a constantly adjusting field, with seasonal changes reflecting shifts in the field's ripple memory.

Key Insight: Seasons are not the result of Earth's motion—they are the consequence of the field's memory unfolding in predictable cycles. Earth's position remains still, but its relationship with the surrounding ripple field shifts, creating the observable cycle we call "seasons."

9.4 Why Earth Feels Still—Because It Is

Why does Earth feel still? Because it *is*. Not symbolically, not metaphorically—physically still within the ripple architecture of Structured Vacuum Energy (SVE). This isn't intuition; it's a reframing of physical reality based on alignment fields, not orbital assumptions. In UED, motion is not required to explain natural cycles. Alignment is.

The Impossibility of Detecting Motion from Within a Synchronized Ripple Shell

If you exist *within* a stable ripple system—where all surrounding elements are perfectly co-aligned—you will not perceive relative motion, because:

- You are embedded in the same reference phase as everything else.
- There is no relative displacement, only field *coherence*.
- What we call "stillness" is simply perfect alignment with no ripple offset.

This is why you feel like you're on solid ground. This is why, despite claims of rotation at 1,600 km/h, you do not feel spun or hurled outward.

Because you're not spinning. You're synchronized.

In UED terms, Earth is not a rotating object—it is a ripple shell stabilized into equilibrium with its local memory imprint. Your experience of stillness is not deception—it's the natural consequence of equilibrium.

Inertia, Coriolis, and Foucault's Pendulum—Reinterpreted

Mainstream science points to inertia, the Coriolis effect, and Foucault's Pendulum as definitive proof of Earth's rotation.

But these phenomena are ripple alignment echoes—not evidence of spin.

Inertia:

- Classically, objects resist change in motion.
- In UED, inertia is not momentum—it is field memory resistance to realignment.
- What we call "mass in motion" is just ripple configuration resisting phase distortion.

Coriolis Effect:

- The deflection of winds and currents on Earth is attributed to its rotation.
- In UED, this deflection arises from asymmetric ripple tension zones, not motion.
- Atmospheric pressure systems and ocean flows align along field curvature, not spinning surfaces.

Foucault's Pendulum:

- Said to demonstrate Earth's rotation as the pendulum's plane slowly rotates
- But the motion is actually due to gradual ripple realignment with Earth's larger field memory, not physical spin.

These effects do occur, but they are not *proofs* of planetary motion. They are consequences of ripple field dynamics and Earth's interaction with the broader SVE.

Earth Doesn't Need to Move to Explain Anything

When you view the universe through ripple dynamics, Earth's motion becomes unnecessary:

- Day/Night cycles? Ripple alignment thresholds.
- Seasons? Cyclic memory tension resolution.
- Gravity? Ripple compression, not planetary pull.
- Tidal forces? Lunar field phase coupling—not orbital gravitation.

Earth sits as a stable ripple node, a central anchor of local equilibrium in a living field of consequence and return. It doesn't spin, rotate, or orbit. It remains fixed while the conditions of alignment change across it—*and that is enough* to create all observable effects.

Key Insight: "The Earth feels still because it is. Motion was our illusion to explain phenomena we didn't yet understand as alignment behavior." You are not on a spinning rock—you are seated within a still but responsive ripple cradle, held by the memory of equilibrium.

9.5 The Atmosphere Isn't a Light Diffuser

We've been taught that the atmosphere scatters sunlight, and that's why the sky is blue and sunsets are red. That the light comes in, hits particles, and bounces around until your eye catches the leftovers. In this model, the atmosphere is a passive diffuser—a cloud of gas molecules redirecting photon traffic.

But under Universal Energy Dynamics (UED), this is an outdated metaphor—a patch to make Newtonian optics and wave theory agree. The truth is more elegant, and more precise:

The atmosphere doesn't scatter light. It resonates it.

The Sky Is Not Blue Because of Scattering

Rayleigh scattering was a clever model. It filled a knowledge gap. It said shorter wavelengths (blue) scatter more due to their size relation with atmospheric molecules. But it makes several problematic assumptions:

- That light is a traveling wave or particle
- That molecules act like billiard balls
- That your eye sees what is left after random collisions

These are visual crutches—illusions made to comfort our classical mindset. In UED, none of these assumptions hold. Light is not a traveling object. Nothing moves. Only fields align.

So why is the sky blue?

Ripple Resonance Across Atmospheric Layers

The atmosphere is a field medium—a multilayered ripple shell around Earth. It's not scattering light—it's tuning your ripple perception.

Here's how:

- The sun's ripple alignment tries to reach your field through the atmospheric shell.
- That shell has density gradients, temperature bands, and pressure zones—each altering the ripple memory coherence.
- Different parts of the spectrum resonate differently through these layers.
- What we perceive as "color" is simply the dominant harmonic of ripple resolution at a particular observer location.

During the day, high-energy (blue) ripple agreements dominate the upper atmosphere's alignment phase.

At sunset, when the ripple angle must pass through more atmospheric density, lower-frequency harmonics (red) achieve better resonance.

So you're not seeing scattered particles—you're witnessing phase-tuned resonance.

Atmosphere as Ripple Shell, Not Photon Trap

Think of the atmosphere like a tuned drumhead, not a cloud of bouncing particles. Its shape, thickness, and structure modulate the ripple field incoming from the sun. Depending on the:

- Observer's location
- Sun's apparent angle
- Atmospheric coherence and humidity
- Field tension and background memory

...different ripple patterns are allowed to resolve in your perception system. That's why twilight looks different from noon, and sunsets are never quite the same twice. It's not random—it's harmonics.

UED Summary

- There are no photons to scatter.
- There is no "beam" of light to be broken.
- There is only Structured Vacuum Energy modulated through atmospheric ripple conditions.
- Your perception of light's color is a field agreement, not a pigment show in the sky.

Key Insight:

"You never saw blue sky or red sunset. You tuned into a ripple symphony your brain was wired to interpret."

9.6 Sun Movement Is Perceived, Not Real

We say the Sun rises and sets. We draw lines across the sky and call them paths. We model orbits, rotations, and celestial mechanics. And we're convinced—utterly convinced—that the Sun is "moving" across our sky.

But it never moved.

In Universal Energy Dynamics (UED), nothing travels. There is no beam, no particle, no light ray flying toward you. There is only ripple alignment. And when you observe the Sun, you are not watching a body shift positions—you are experiencing a changing field resonance.

The Sun Doesn't Rise. You Realign.

The illusion of solar motion emerges from a simple yet profound misunderstanding:

We thought we were watching motion. We were actually watching a ripple knot pass through our perception field.

Here's what's actually happening:

- The Sun's ripple consequence—its field echo—is always active, extending into Structured Vacuum Energy (SVE).
- Your body, your location, your atmosphere—all represent a ripple node within SVE, tuned to a certain perceptual threshold.
- As global ripple alignment shifts (due to memory tension and field cycles), your position intersects the alignment window that creates the illusion of "seeing the Sun."

That's sunrise. That's sunset. Not the Sun rising or setting—but your ripple matching condition appearing, peaking, and fading.

No Light Was Sent—Only Agreement Achieved

Mainstream science says:

Light left the Sun 8 minutes ago and just reached your eye.

UED says: It took 8 minutes for the ripple field between the Sun and your eye to synchronize. No particle traveled. The delay is not motion—it's Alignment Threshold Time (ATT).

"Solar Motion" Is Ripple Knot Traversal

Imagine a ripple knot—a localized region in SVE where the Sun's ripple consequence coheres best with your Earth field. As

Earth's local field memory shifts (due to tension cycles, atmosphere dynamics, and ripple discharge history), the knot appears to move.

But the Sun? It stayed where it was. It never sent anything. It never moved.

What moved is your point of optimal alignment.

You didn't track the Sun. You tracked your changing ripple agreement.

UED Summary:

- The Sun's "movement" is an illusion born from ripple re-alignment.
- No light travels. No object moves. Only alignment knots migrate through the observer's perceptual threshold.
- Your experience of sunrise and sunset is not spatial—it's field-relative alignment unfolding.

Final Insight:

"You didn't watch the Sun cross the sky. You felt your ripple field catch, hold, and release the echo."

9.7 What Stars and Planets Really Are

For thousands of years, we've assigned stars their glittering roles as fiery spheres and planets as orbiting neighbors in a cosmic dance. Telescopes gave them surface features. Physics gave them mass. Mathematics gave them paths.

But none of these are real entities in the way we thought.

Universal Energy Dynamics (UED) offers a radical reinterpretation:

Stars and planets are not objects flying through space. They are ripple phenomena—structured alignments and memory tensions within the Structured Vacuum Energy (SVE).

Let's dismantle this sky—one layer at a time.

Stars Are Not Balls of Fire

The common belief: Stars are distant suns—giant thermonuclear furnaces burning light into space.

UED says: Stars are remote ripple emitters. They don't "shine" by throwing photons across space. Rather, they represent long-distance ripple tension nodes that interact with the SVE grid to create observable consequences—not light, but field resonance.

You don't "see" a star because its light traveled to you. You "see" it because your local field resonates with its distant ripple signature.

Even when a star is said to be dead for millions of years, yet still "visible"—that's not a delay. It's a persisting ripple alignment memory, not a flying photon echo.

"Starlight" is a present field handshake, not a past-tense postcard.

Planets Are Not Flying Spheres

In current models, planets are solid objects caught in orbit due to gravitational pull. In UED, planets are ripple nodes—local field concentrations that:

- Maintain long-term tension balance with SVE
- Appear to "orbit" due to observer-relative ripple overlays
- Don't revolve—they cohere

A planet is not a traveler. It is a persistent ripple knot stabilized by surrounding alignment conditions. Its shape, motion, mass, and gravitational behavior are all field consequences not core properties.

Retrograde Motion and Eclipses: No Mysteries—Just Ripple Overlaps

Retrograde motion—where a planet appears to move backward—isn't a mechanical illusion of orbits. It's a temporary interference pattern where the ripple of one planet crosses the alignment path of another, altering the visual resonance for the observer.

Eclipses? Field occlusion. Not shadow play.

- A solar eclipse is ripple phase blockage, not sunlight blocked.
- A lunar eclipse is temporary phase noise, not literal shadow falling.

All these are consequence effects in ripple choreography. No one body has to "move" or "hide." It's just alignment disruption.

What This Means

- The stars don't shine—they resonate.
- The planets don't move—they persist as ripple harmonics.
- Their "paths" are alignment overlays—field music, not cosmic dance.

Key UED Reframes:

Phenomenon	Classical View	UED Interpretation
Starshine	Photons from nuclear fusion	Ripple resonance from a distant emitter
Planet orbits	Elliptical movement due to gravity	Stable ripple knots with shifting alignment
Retrograde motion	Apparent reversal due to relative motion	Observer-field overlay altering alignment
Eclipses	Shadow effects	Phase tension obstructions in SVE
Visibility	Light arrival	Ripple field agreement

Final Insight:

"Stars never arrived. Planets never left. The sky isn't a map of movers—it's a memory board of alignments."

9.8 Closing: The Ground Was Never Lost

We were told that Earth is spinning. We were told it's hurtling through space. We were told we're passengers on a rock, locked in cosmic inertia.

But all of it—every turn, every orbit, every revolution—was perception. Not fact.

Universal Energy Dynamics (UED) rewrites the closing argument:

Earth is not moving. It is remembering.

Earth: The Ripple Foundation Node

At the heart of all experience lies a stillness we forgot to believe. Earth is not a projectile. It is a structural knot of ripple coherence in the Structured Vacuum Energy (SVE) grid.

It holds ripple memory, receives alignment pressure, and manages consequence flow. It feels still because it is.

Everything that seems to move above it—the stars, the planets, the Sun—is not motion, but a shifting agreement pattern in the surrounding field. A dance of ripple re-alignments, not object trajectories.

You don't spin into day. You tune into dawn.

You don't revolve into summer. You resonate with a cycle.

Liberation from Illusion

This is not an attack on science. It's a release from patchwork physics built on assumptions of motion, force, orbit, and shadow.

Stillness is not stagnation. Stillness is the precondition for meaningful change.

Without stillness, alignment has no meaning. Without a foundation, no perception could form.

"The Earth didn't carry you through space. It held you still enough to notice the universe ripple around you."

The True Ground of Being

This chapter concludes with a truth many won't accept easily:

- We didn't need orbital mechanics to explain seasons.
- We didn't need planetary spin to explain day and night.
- We didn't need to abandon the ground to understand the sky.

We needed to abandon the illusion that motion = meaning.

In UED, we don't chase reality—we witness consequence emerge from aligned stillness.

Final Thought:

"The ground was never lost. We were simply too dazzled by motion to feel it under our feet."

What We Just Shattered

"You Never Moved—You Remembered Differently"

Let's distill what this chapter dismantled—gently, but without apology:

The spinning Earth myth: We were taught Earth rotates to create day and night. But under UED, no such motion is needed. What changes is the ripple agreement—not the planet's position, but the observer's alignment with the ripple field.

The orbital seasons illusion: Seasons don't require a revolving Earth. They're cyclical field phenomena, emerging from ripple memory discharges that pass over Earth like tides through tensioned space.

Sky color, twilight, and sunrises explained as scattering: We were told the atmosphere diffuses light. But scattering is a patch. The atmosphere resonates with ripple frequencies, tuning visual perception—not breaking light apart.

Motion as proof of truth: We shattered the obsession with movement as a requirement for cause. Stillness is not denial. It is structure.

Celestial motion: Stars and planets don't orbit in the ways we assumed. Retrogrades, eclipses, night skies—they are ripple overlays, not object migrations. The sun never moves. You do—through your perception.

UED's Gift in This Chapter:

Earth didn't spin to show you the Sun. It stayed still so you could witness the ripple alignment.

We didn't dethrone Earth—we restored its dignity. We didn't reject astronomy—we gave it a new pulse.

The cosmos hasn't been flung outward—it's been vibrating in place.

"You never moved. You simply remembered differently, each time the field realigned."

Interlude: The Dust Begins to Settle

"You Left Illusion. Now Welcome to the Real."

In Part I, we stripped away what was never solid.

Particles, motion, force, time, and spacetime—each one fell not because it was weak, but because it was never truly there.

But destruction without reconstruction is nihilism.

You don't just break the temple of illusion—you build something better in its place. That's exactly what Part II offered: a new lens to see the universe not as a dance of objects, but as the choreography of consequence.

What Just Happened?

- We introduced Universal Energy Dynamics (UED)—a framework not of entities and forces, but of ripples, delays, alignments, and field memories. Not energy acting—but energy remembering.
- We removed light from the pedestal of travel. No photons flew from stars into your eyes. You never saw the Sun—you aligned with its ripple echo.
- We demystified force itself—magnetism, gravity, radiation—none are things. They're behaviors. Field moods. Memory tensions.
- And finally, we reclaimed Earth's dignity. No more spinning rock hurtling through space. Earth is not in motion—it is the still node, around which the illusion of movement

dances through ripple alignment.

Why This Part Matters

Because once you let go of force, source, emission, and motion, the world doesn't fall apart.

It becomes simpler.

The world becomes less crowded. Less chaotic. Less about pushing and pulling. More about tuning, aligning, remembering.

As We Step into Part III...

You're not merely learning new physics. You're shedding inherited illusions, replacing what perception insisted on with what existence always encoded—quietly, patiently, beneath the noise of equations and theories.

"You were never pushed. You aligned."

Now, we turn toward the core machinery of science itself—quantum mechanics, relativity, and the sprawling speculative architectures built to patch their holes.

We're not here to rebel. We're here to disentangle misunderstanding from mystery.

Part III begins. The collapse of the accepted will now become the rise of the coherent.

Part III: The Systemic Collapse

Chapter 10: Quantum Patchwork

Why Duality Was Just a Ripple Misread

"Reality is not uncertain. Our instruments—and assumptions—were."

For over a century, quantum physics has dazzled the world not because it made sense, but because it *worked* despite making none. Wave-particle duality, superposition, collapse of the wavefunction—these weren't elegant explanations. They were desperate metaphors trying to hold together a leaking framework.

In a world where electrons behave like particles *and* waves, and where a cat is both alive and dead until you peek, the mind of the physicist became not a lens—but a crutch. Measurement wasn't explanation; it was a surrender. Quantum theory didn't illuminate reality—it dodged it.

This chapter is not here to ridicule quantum physics. It honors its predictive power. But it also exposes its cracks—not by rejecting its data, but by reinterpreting it through the clearer lens of Universal Energy Dynamics (UED).

Here, we replace *duality* with *ripple memory*. We replace *observer weirdness* with *threshold alignment*. We discard the illusion of collapse and embrace a universe that always knew what it was doing—even when we didn't.

Lets deconstruct the foundational contradictions in quantum mechanics—especially wave-particle duality and observer-dependent phenomena—and show how Universal Energy Dynamics (UED) resolves them as misunderstood ripple behaviors and alignment thresholds within the Structured Vacuum Energy (SVE) field.

10.1 The Crisis Quantum Never Solved

"Quantum theory predicted brilliantly. But it explained nothing."

The early 20th century was a chaotic awakening. As classical mechanics began to fail in describing phenomena at atomic scales, physicists like Planck, Einstein, and Bohr cracked open a door—only to find themselves staring at a hall of mirrors.

Planck's constant was not a revelation—it was a rescue rope. He introduced quantization not because it made sense, but because blackbody radiation wouldn't behave otherwise. Einstein extended this patchwork with the photoelectric effect—not because light was suddenly discovered to be a particle, but because existing models couldn't explain electron ejection without imagining "quanta" of light. Bohr then tiptoed across a razor's edge by postulating fixed orbits for electrons—not derived from logic, but forced by spectral data.

And thus, quantum theory was born—not as clarity, but as accommodation.

With each new observation that classical physics couldn't explain, a quantum rule was added—superposition, quantization, uncertainty, duality. These weren't interconnected principles springing from a unified worldview. They were ad hoc plugs for systemic leaks. Physics became less about "what is" and more about "what works." Prediction ruled over understanding.

It was in this crucible of contradiction that physicists made a pact with pragmatism:

"If the math works, don't ask what it means."

But cracks remained—deep, conceptual ones. How could something be both wave and particle? Why did an observer affect the outcome? What was "collapsing"? How could a particle "exist" in multiple places at once?

Quantum mechanics refused to answer. It only pointed to results.

And because those results matched experiments—every paradox was forgiven. Quantum became a religion with working miracles—but no explanation for the divine.

Universal Energy Dynamics (UED) enters not to destroy quantum physics, but to decode it. To reveal that what quantum theory described awkwardly through dualities and uncertainties were actually ripple behaviors in a structured field. Not particles in magic states—but ripple alignments awaiting threshold fulfillment.

In UED, we won't discard quantum data. We will reinterpret it with elegance.

We're not fixing what's broken—we're explaining why it never made sense to begin with.

10.2 Duality: The Biggest Misinterpretation

"You weren't seeing duality. You were switching lenses mid-consequence."

Wave–particle duality is perhaps the most celebrated contradiction in all of modern physics—and the most misunderstood.

We were told an electron can act like a particle and a wave. A photon can behave like a pinpoint impact or an interference ripple. Depending on how you looked at it—through which apparatus, under what conditions—the very nature of reality seemed to shift.

This was hailed not as a red flag, but as a mysterious feature of the quantum world. A paradox that could not be resolved—but must simply be accepted.

And yet, there was no duality. There was only confusion born of limited perception.

UED reveals that what we labeled as "wave" and "particle" are just two different consequence profiles of a single ripple behavior unfolding in the Structured Vacuum Energy (SVE) field.

When the ripple is observed before alignment stabilizes, we see interference, superposition, fuzziness—what we call "wave" behavior. When we measure the ripple after consequence localization, we detect impact, quantization, path—what we call "particle" behavior.

But the ripple was never two things. It was never switching modes.

We were.

We built machines that interfered with the ripple's coherence. We set up detectors that collapsed consequences by forcing alignment. And then, bewildered, we blamed the ripple for our own shift in perspective.

Think of a shadow on a wall: is it sharp or fuzzy? That depends on the angle of light, the shape of the object, the surface texture. It is one phenomenon, but perceived differently based on your frame of reference. We don't say the shadow is in dual states. We say our viewing conditions changed.

Similarly, particles never acted like waves. They were always ripple consequences contextualized by how we engaged with them. It was not duality—it was lens-switching mid-event.

The truth is sharper than paradox. Duality was a patch to explain why our instruments contradicted themselves. UED doesn't need such a patch. It recognizes that ripple alignment behaves differently across thresholds of coherence, and the outcomes are observer-aligned consequences, not mysterious shapeshifting entities.

Let's stop mythologizing confusion. There is no quantum weirdness—only ripple logic misread through classical assumptions.

10.3 Double Slit: The Ripple Rewritten

"It was never a photon walking two paths—it was your field catching echoes of ripple negotiation."

No experiment is quoted more often in quantum mysticism than the double-slit experiment. And no experiment has been more misrepresented.

In the traditional narrative: A photon or electron is fired toward a barrier with two slits. When no detector observes the slits, interference appears—suggesting wave behavior. When a detector is placed to watch which slit the particle goes through, the interference vanishes—and particle behavior is observed.

The conclusion? Somehow, the photon "knows" it's being watched. It "chooses" to go through one slit or both, to behave like a wave or a particle based on observation. This birthed a century of poetic confusion: "Consciousness collapses the wave function." "Reality doesn't exist until you look."

But Universal Energy Dynamics (UED) clears this fog decisively.

There was no traveling photon. No object flitting through space and making decisions. What the experiment measured was not the movement of a particle, but the field alignment consequence at the detector's surface. The ripple didn't "choose" a slit—it configured a resonance path through the structured vacuum field.

Here's what really happened:

The emitter introduced a localized ripple into the SVE (Structured Vacuum Energy).

The slit barrier modulated the ripple—forcing reconfiguration as it passed through zones of constraint.

This modulation didn't split the photon—it created ripple interference tension within the field.

The detector didn't record a traveling particle. It recorded the alignment threshold where ripple reinforcement (or cancellation) occurred—what we call the interference pattern.

Now add a which-path detector, and everything changes—not because the photon "knows," but because the detector itself disturbs the ripple field, collapsing the coherence before the field can complete its negotiation. The result? No interference—because no complete ripple realignment.

The interference was never "mystery behavior." It was ripple choreography across misalignment boundaries, trying to find harmony within the SVE. The experiment did not reveal duality—it revealed the fragility of ripple coherence when observed without awareness of field dynamics.

The double-slit experiment was not proof of quantum weirdness. It was proof that we never understood the medium. We saw patterns emerge and assumed particles walked paths.

In truth, you never saw a photon. You witnessed the ripple field's agreement with your device's alignment. Not a thing moving. Just a consequence appearing where conditions matched.

10.4 The Observer Effect Demystified

"The observer didn't collapse a wave. They completed an overdue handshake."

The observer effect has long haunted the halls of quantum physics. It has inspired both awe and absurdity—from "consciousness creates reality" to "nothing exists unless it is observed." But what if the mystery isn't in the observer... ...but in how late we understood the ripple alignment conditions?

Universal Energy Dynamics (UED) reframes the so-called "observer effect" not as an act of magical measurement or conscious intervention, but as a ripple resonance event. Here's how:

In UED, a ripple field exists in a state of negotiation. It oscillates, aligns, resists, or diffuses depending on local ripple pressure and structural memory. Observation—or what mainstream physics calls measurement—is nothing more than a ripple alignment threshold being crossed between two field systems: the emitter and the receiver (including you).

There's no "collapse" in the wave function. Nothing dissolves into a definitive state. There's only a lock-in—a finalization of ripple behavior due to coherent boundary contact.

That boundary? You. Or more precisely, your observational tool's field and its minimum requirement for detection.

This is where UED introduces a key concept: ATT – Alignment Threshold Time.

Every ripple—whether we call it "light," "charge," or "wave"—takes time to align with a receiver's ripple configuration. It is not instantaneous. It is not purely probabilistic. It's a consequence of field readiness—a duration during which the ripple tries to find enough symmetry to resolve as a perceptible event.

If it fails? No detection. If it succeeds? We call it "measurement."

But the ripple was always there—waiting, negotiating, aligning. Not a ghost in a box. Not a dual-natured particle choosing its mask. Just a consequence delayed until the handshake could be completed.

This fundamentally rewrites the observer effect:

The observer does not influence an outcome by looking.

The observer becomes the final field needed for the consequence to emerge.

The "collapse" is merely the end of ripple ambiguity, not because of intent—but because of alignment completion.

No mysticism. No "does the Moon exist when you're not looking?" paradox.

The Moon exists—but not as an object. It exists as ongoing ripple consequence—independent, but always waiting for you to align, so it can finish its ripple expression in your field.

The observer did not create reality. They synchronized with it—just in time.

10.5 Superposition and Entanglement: Misread Memories

"What you called spooky action... was just memory alignment reaching its overdue equilibrium."

Quantum mechanics gave us two of the most baffling concepts in physics: Superposition—the idea that a system can exist in multiple states until measured. And Entanglement—the idea that two particles, separated by vast distances, can instantly affect each other.

But these are not phenomena. They are symptoms—of a deeper misunderstanding of ripple dynamics. Let's demystify both through the lens of Universal Energy Dynamics (UED).

Superposition:

In quantum language, superposition is the state of a system before a measurement—a blur of possibilities. But under UED, there is no blur. No contradiction. No particle "being in two places."

Superposition is nothing more than an unresolved ripple alignment. The ripple field hasn't finalized its interaction yet—it's in negotiation. The ripple isn't undecided—it's waiting for sufficient resonance from the observer's field to resolve itself.

You think the cat is alive and dead. UED says: the ripple state has not aligned with your field condition—so no outcome has finalized. You aren't witnessing dual existence. You're witnessing delayed agreement.

There's no mystery. No paradox. Just ripple patience.

Entanglement:

Now the crown jewel of "weirdness"—entanglement.

In standard quantum mechanics, two particles interact, then fly apart, and somehow retain the power to "know" each other's state—even across galaxies.

Einstein called it "spooky action at a distance." UED calls it: compressed ripple memory across fields.

When two systems interact, their ripple histories become correlated. They share an imprint—a pattern of ripple alignment that remains embedded in the Structured Vacuum Energy (SVE).

When you "measure" one, you're not sending information across space. You're accessing the shared ripple alignment record, embedded in the same universal substrate.

There's no signal. There's no need for "instant communication." It's like two ripples in a pond being born from the same drop—they're still consequences of a single event. When one completes a pattern, the field rebalances itself across all nodes involved.

Entanglement, in UED, is not a connection between particles. It's a connection between field expectations—a synchronized ripple pressure held in equilibrium memory.

What seems like instant change is not action—it's pre-agreed resonance unfolding once enough alignment pressure is reached.

In other words:

- You didn't flip the state of the distant particle.
- You restored the alignment condition between two historically entangled zones.
- The universe didn't "send" anything—it simply let the tension go.

Superposition is delay. Entanglement is memory. Both are ripple-based phenomena, misinterpreted by mathematics and mystified by language.

What physics called "weird" was simply structured memory trying to settle itself—while we kept measuring it too early or from the wrong frame.

10.6 Probability Is Ripple Ignorance

"What you called chance was just unread ripple consequence."

Quantum mechanics is built on a throne of probability. The wavefunction—ψ—isn't just a tool in standard theory. It's the very altar at which quantum physics prays. It doesn't predict what *will* happen. It tells you the *likelihood* of outcomes.

But UED breaks that illusion.

The Wavefunction Is Not Reality

Let's begin with the myth: Physics declared, "We can't know what a particle *is*—only what it *might do*." So they replaced understanding with statistics, and called it fundamental. But in Universal Energy Dynamics (UED), the wavefunction is not the fingerprint of reality—it is the fog over consequence.

The wavefunction represents unresolved ripple conditions. It's a shadow of what we haven't aligned with, not a map of nature's true intention.

Probability: The Mask of Misalignment

Where does quantum probability arise from?

Not from nature being uncertain. Not from particles flipping coins. But from our incomplete knowledge of ripple field conditions.

Every "random" quantum event—whether in decay, spin, or position—is a reflection of our ignorance of underlying ripple tension.

We don't calculate probabilities because nature is gambling.

We calculate them because we can't yet see the full ripple memory structure beneath the event.

In UED terms:

- A 50% chance of spin up isn't randomness.
- It's a field split not yet resolved through alignment.
- When the ripple consequence meets your observer field and crosses the Alignment Threshold Time (ATT), the

outcome finalizes—not randomly, but necessarily.

Causality Isn't Dead. It's Hidden in the Ripples.

Quantum mechanics gave up on causality. UED restores it—not through determinism, but through structured memory and ripple logic.

In UED:

- Every outcome is a field response—not a roll of the dice.
- Every pattern is consequence-driven, not chance-led.
- Every "random" event is just a ripple sequence we failed to decode in time.

Probability is not a law. It is a placeholder for ignorance.

So what does UED say to quantum indeterminacy?

"You were not watching nature being uncertain. You were watching yourself being unaligned."

10.7 Quantum Tunneling and Nonlocality—Reframed

"Nothing skipped the wall. It realigned beneath your assumptions."

Two of the most misunderstood phenomena in quantum mechanics—tunneling and nonlocality—have baffled physicists for over a century. They sound magical. Mysterious. Borderline mystical. Particles "teleport" through barriers. Entangled twins "communicate" across galaxies. Einstein even mocked it: *"Spooky action at a distance."*

But UED sees something far more elegant—and far less spooky.

Quantum Tunneling: Not a Violation—A Ripple Shortcut

In traditional QM, tunneling implies that a particle somehow passes through a barrier it doesn't have the energy to cross. An electron in a potential well, for example, "leaks" through the wall. No explanation—just math that works.

UED reinterprets this event without invoking teleportation or absurd probabilities.

In UED terms:

- The particle never moved.
- The ripple field on both sides of the barrier remained in phase tension.
- The ripple node realigned to the other side when the SVE (Structured Vacuum Energy) phase conditions across the barrier reached compliant resonance.
- What seemed like a leap was actually a ripple compression collapse—a field shortcut, not a path crossing.

"Tunneling is not matter breaking barriers—it is field relieving pressure through phase memory."

Nonlocality: A Ripple Memory Echo

Entangled particles are the poster children of quantum weirdness. You measure the spin of one—and instantly know the spin of the other, no matter how far away.

Standard QM says: *It just works.* Some say it's nonlocal information transfer, others say it's retrocausality.

UED says: *There is no "transfer."* There is only shared ripple memory.

When two ripple zones emerge from the same compression event (such as a pair production), they inherit a shared memory imprint in the SVE field. This imprint is non-geometric—it's a pattern of consequence that extends across the ripple topology, not physical space.

So when you "observe" one, you're not sending information. You're simply realigning a part of the memory that forces compliance in the entire imprint.

No signal. No speed. No paradox.

Field Knots vs. Particle Paths

In UED, what we call a "particle" isn't a point—it's a knot of ripple alignment. That knot doesn't *move* across space. It relocates when the pressure to maintain alignment becomes too high in one region and too low in another.

That's why tunneling can seem instantaneous. And entanglement can appear nonlocal. But in both cases—nothing is travelling. What's happening is memory syncing and ripple logic restoring equilibrium.

Summary Thought:

"Quantum mechanics called it magic. Universal Energy Dynamics calls it inevitability—the field reconfiguring itself in the only way it can: without distance, without delay, without doubt."

10.8 Closing: Quantum Was the Language of Confusion

"Quantum theory was never wrong. It was just never whole."

Quantum mechanics didn't fail us—it saved us from total classical collapse. But it did so the way a splint saves a broken leg: temporarily, awkwardly, and with side effects.

It delivered staggering predictive power—lasers, transistors, MRIs, semiconductors—but behind this power lay a sea of contradictions, dualities, and paradoxes that were quietly swept under the theoretical rug.

Clarity Traded for Calculability

The genius of quantum mechanics was not in its ontology, but in its calculus. It said: *We don't need to know what's "really" going on—just give me the probabilities.* And for decades, that bargain worked.

But here's what was lost in the transaction:

- What is a particle, really?
- Why does observation "collapse" a wave?
- How can something be in two states at once?

- Why does entanglement happen instantly?
- What does it mean to measure something?

None of these questions were answered—only redirected into equations, backed by consensus, and defended by convention.

The Quantum Patch Became the Religion

Physicists began worshiping the mystery. Some called it "weirdness," others "non-intuitive beauty." But at its core, quantum mechanics became a language of confusion accepted as doctrine—because its numbers worked.

Yet any theory that thrives on paradox, contradiction, and denial of causality cannot be the final word. It can only be a transitional tongue, pointing toward something deeper, more elegant, and ultimately more real.

What UED Offers Instead

Universal Energy Dynamics doesn't dismiss quantum mechanics. It absorbs it—and then reorganizes it. The dualities, entanglements, uncertainties, and tunnels are all preserved—but reframed.

Instead of:

- Wave vs particle
- Randomness vs determinism
- Collapse vs continuity

UED offers:

- Ripple alignment
- Structured memory
- Threshold-driven perception

No magic. No mystery. Just a single coherent substrate—Structured Vacuum Energy—governed by consequence and ripple pressure.

Final Thought:

"Quantum physics whispered a broken truth in a beautiful language. UED listens deeply—and translates it into meaning."

What We Just Untangled

"You Didn't Collapse the Wave—You Misread the Ripple."

Wave–Particle Duality

What they thought: A particle behaves like a wave until measured. What we clarified: You were perceiving ripple consequences through shifting alignment—not switching identities mid-flight.

"The duality was never in the thing. It was in your lens."

The Observer Effect

What they thought: Measurement causes a wavefunction to collapse. What we clarified: Measurement is just a threshold of ripple interaction. The field doesn't collapse—it resolves.

"You didn't cause a change. You reached the point of visible agreement."

Double-Slit Experiment

What they thought: Particles go through both slits simultaneously. What we clarified: Interference arises from ripple tension resolving across potential alignments—no particle ever traveled.

"Patterns formed not by paths taken, but by possibilities rippling."

Entanglement and Superposition

What they thought: Particles share mysterious, faster-than-light connections. What we clarified: Ripple memory is shared across fields. Entanglement is not magic—it's synchronized compression.

"Entangled states are not distant twins—they're two faces of one ripple memory."

Tunneling and Nonlocality

What they thought: Particles defy barriers and jump instantaneously. What we clarified: Ripple compliance across compressed zones creates seamless continuity—not teleportation.

"What seemed sudden was just ripple resolution bypassing your logic."

Probability and Uncertainty

What they thought: Reality is fundamentally random. What we clarified: Uncertainty is not a limit of nature—but of interpretation. Probability emerges from ignorance of ripple alignment, not inherent chaos.

"You calculated likelihoods because you never mapped the ripple field."

Summary:

Quantum theory wasn't wrong—but it was incomplete. It patched over mysteries with mathematics, delivering results without explaining origin. UED doesn't discard those results. It grounds them in ripple logic, structured memory, and consequence-first physics.

"You didn't observe a mystery. You observed a message—garbled by an outdated lens."

Now, as we step into the collapse of relativity, let's keep untangling the myths that equations protected—but truth never needed.

Chapter 11: Relativity Unraveled

What Einstein Got Right—and What We Had to Fix

Let's now demystify the most celebrated concepts of Special and General Relativity—without discarding their insights, but reframing them through Universal Energy Dynamics (UED). We'll show that time dilation, length contraction, and even black holes are not distortions of spacetime—but ripple field consequences, shaped by alignment, delay, and compression in Structured Vacuum Energy (SVE).

He bent space. He curved time. He taught us to think in four dimensions and convinced us that gravity wasn't a force—it was the shape of the universe itself.

Einstein didn't just rewrite Newton. He reimagined reality. And for a century, we've revered his equations as gospel—because they worked. Satellites sync with relativistic precision. GPS clocks account for time dilation. Black holes were predicted before they were "seen."

But what if we told you: he saw the right shadows on the wall—but misread the flame that cast them?

Relativity was brilliant. But it bent the map to fit the landscape. UED does the opposite—it shows the ripple logic behind the terrain itself.

This chapter isn't a rejection of Einstein. It's a rescue mission—to preserve his insights while freeing them from the geometric illusions they were forced into.

Because space never bent. Time never stretched. The field simply aligned—and we mistook it for curvature.

Let's unravel the tapestry—and trace the ripple that was always behind the thread.

11.1 The Relativity Revolution

The 20th century didn't just change physics—it redefined reality. At the center of this transformation stood a man who replaced gravitational pull with curvature, clocks with coordinates, and absolutes with relativity.

From Newton to Einstein, the shift wasn't just scientific—it was philosophical.

Newton gave us a universe of invisible forces: gravity as an attractive thread stretching across void. Time ticked uniformly, space stayed still, and objects danced in a stage set by absolutes. His equations worked wonders, from apple drops to planetary motion—but they couldn't explain the underlying mechanism. Action-at-a-distance remained a ghost.

Then came Einstein.

In Special Relativity, he dissolved simultaneity and rebuilt reality on the constant speed of light. Time became stretchable, length became contractible, and observers became central to the measurement of everything. Space and time were no longer separate—they were married into a four-dimensional continuum.

In General Relativity, he took a deeper leap: Gravity wasn't a force—it was geometry. Massive objects "curved" spacetime, and free-falling bodies simply followed the new contours. Planets didn't orbit—they rolled along invisible bends in the fabric of existence.

This was the revolution. And it worked. Mathematically, stunningly, experimentally—it worked.

So well, in fact, that we stopped questioning *why*. Why do massive bodies curve spacetime? Why does time slow near speed or mass? Why should light "bend" if there's no medium, no wave, and no force?

These were set aside—sacrificed at the altar of prediction. Because relativity predicted, it was crowned as truth.

But UED—Universal Energy Dynamics—offers a revision: not a denial of Einstein's accuracy, but of his framing. We say: Geometry was the metaphor. Ripple alignment was the mechanism.

Einstein gave us a brilliant map. But we mistook that map for the terrain. What he modeled as curvature was actually ripple compression. What he called time dilation was ripple delay. What appeared as gravitational lensing was a memory path detour in SVE.

Relativity was a revolution—but it never explained what rippled, what aligned, or what remembered. It told us how things appear when energy patterns shift, but not why they shift at all.

Now we look again—not to dismantle the past, but to reveal its missing foundation.

Because space never bent. It remembered.

11.2 Time Dilation: A Ripple Delay

Relativity told us something strange: time can stretch. That a moving clock ticks more slowly. That gravity makes your wristwatch crawl compared to one in deep space. That two twins, one on Earth and one on a spaceship, will age differently.

And yet—none of it felt *real.*

Because we never saw time stretch. We saw clocks desynchronize. We saw processes delay. And we called it time dilation.

But what really happened?

Let's strip away the spacetime metaphor and look at reality through the UED lens:

Time isn't a dimension. It isn't a fabric. It isn't a thing that can "flow" or "dilate."

Time is a field delay. A lag in ripple consequence. A pause in alignment handover across Structured Vacuum Energy (SVE).

No Clock Ever Slows Down

In UED, the ticking of a clock isn't governed by "proper time" or relativistic contraction—it is governed by ripple alignment rate. Each tick is the system's return to equilibrium, a ripple handshake that completes one cycle of reality's rhythm.

When a system moves rapidly, its local ripples become phase-shifted relative to the surrounding SVE. They struggle to resynchronize with the larger field. This increases the Alignment Threshold Time (ATT)—the minimal delay needed to complete perceptual coherence.

So clocks don't tick slower because time itself stretched. They tick slower because alignment got harder.

It's not time that was warped—it was the effort required to agree.

The Twin Paradox—Demystified

In standard relativity, the twin who travels at high speed and returns home is younger. Time, they say, passed differently for them.

In UED, we say: nothing mystical occurred.

The traveling twin encountered ripple misalignment due to high velocity and field compression. Their ATT increased, meaning all processes—biological, chemical, mechanical—took longer to realign with SVE. Their perception was smooth, but their consequence accumulation lagged behind.

Aging didn't slow because time bent—it slowed because consequence took longer to propagate.

The twin didn't age less. They aligned less.

Time: A Consequence of Delay—Not a Driver

UED rejects the notion of time as a container or coordinate. Time is not where events happen—events are when alignment occurs. What we measure as "time" is actually a count of completed ripple alignments in a given frame.

Delay that alignment, and time seems to stretch. Speed it up, and time seems to shrink.

But in reality, there's no stretching or shrinking. Only delayed echo. Only phase response lag. Only consequence—taking the long way home.

"You didn't slow time—you lagged in consequence."

This is not a denial of what Einstein saw. It's a retranslation of why he saw it.

Time dilation was real. But it wasn't dilation of time—It was dilation of ripple response.

Welcome to a universe where your seconds are shaped not by speed, but by how hard it is to agree with reality.

11.3 Length Contraction: A Misread of Ripple Density

Einstein's Special Relativity told us something unsettling: Objects contract in length when they move fast.

The faster you go, the shorter you become—at least to someone watching. But you don't *feel* shorter. Nothing *breaks* or *bends*. It's all perception, they say.

And therein lies the clue. Because perception, not deformation, is what's truly shifting.

What Relativity Saw—But Couldn't Explain

Length contraction in relativity is derived from Lorentz transformations, which are mathematical tools used to adjust space and time coordinates between moving observers. They work. They predict experimental results. But they don't explain the *why* beneath the *what*.

Relativity says: the object got shorter. UED says: the ripple field got denser.

And you—the observer—aligned with a compressed ripple zone.

Ripple Zones and Perception Frames

In UED, an object is not a thing—it is a standing ripple structure in Structured Vacuum Energy (SVE). Its boundaries, properties, and shape emerge from stable ripple alignments.

Now suppose this structure begins to move rapidly relative to the field—or more precisely, undergoes a phase shift in ripple terms. As ripple density builds up ahead of the motion and stretches behind, the zone of stable alignment tightens in the direction of motion.

To an observer locked into a different ripple alignment (a different frame), this denser zone appears contracted—not because the object shrunk, but because the area of aligned interaction compressed.

You didn't see a shorter object. You encountered a narrower window of ripple coherence.

Lorentz Transformations: Mathematical Ripples

When Lorentz equations adjust length for velocity, they are actually compensating for the way ripple consequence is experienced across different reference alignments. They are not physical deformation maps. They are projection recalibrations—telling you how much of the ripple pattern you're accessing under shifted phase conditions.

Imagine looking at a ripple grid through a tilted lens. The pattern hasn't changed. Your window onto it has.

So you measure less—not because the object is less—But because you are more misaligned.

Restoring Physical Intuition

Under UED:

- A ruler on a fast-moving spaceship isn't compressed.
- Its ripple imprint into the SVE shifts, and only partial alignment is possible from an external field perspective.
- The result? The same object appears contracted to one observer and unchanged to another—not because reality

bent, but because field perspective diverged.

There's no mystery. No paradox. No magic.
Only ripple realignment and perceptual consequence.
"You didn't measure less. You aligned with more."
This is the heart of it:

- Your frame didn't witness a shorter object.
- It interacted with a compressed zone of ripple agreement.

The ruler was not squeezed by velocity. You squeezed the observation through your ripple lens.

Length contraction isn't a physical change. It's a shift in observable tension across ripple densities.

11.4 Gravity Without Geometry

Why Curves Never Mattered—And Compression Did

The apple didn't fall because Earth bent space. It fell because the ripple field beneath it was compressed—and the apple aligned with the easiest path toward ripple relief.

Einstein's General Relativity replaced Newton's invisible "force" with an elegant metaphor: spacetime curvature. Mass bends spacetime, and objects follow those curves.

But what is "spacetime," really? Can you bend a coordinate? Can you curve a grid?

No. Because nothing physical is curving. Something energetic is compressing. And that *something* is the Structured Vacuum Energy (SVE).

UED View: No Curves—Just Compression

In Universal Energy Dynamics (UED), there is no need for an abstract 4D spacetime fabric. Instead, we have SVE—a structured but tensionless background field.

When mass accumulates or organizes, it doesn't "bend space." It compresses the ripple field in its vicinity, distorting the local alignment balance.

This creates gradient tension across ripple zones—And other ripples, structures, or objects naturally realign toward lower tension states.

This realignment is what we call gravity.

Gravity Is Not a Pull—It's Ripple Memory Rebalancing

There is no invisible hand pulling objects down. There is only the field saying:

"You don't belong in this misaligned state. Come closer—here, the ripple fits better."

Gravity is the urge to align, driven by memory compression in the field. The apple doesn't fall because of Earth's "pull." It realigns downward because Earth's field memory is densest below.

This ripple compression builds pressures that objects resolve by moving into better-aligned configurations.

Why Acceleration Feels Like Gravity

General Relativity tells us that acceleration and gravity are equivalent—and this has baffled many thinkers.

But under UED, this equivalence is not strange at all. Because both phenomena involve the same ripple disturbance logic.

- Acceleration creates a local field lag—a misalignment across your own structure.
- Gravity is already a standing ripple compression—pulling you into equilibrium.

In both cases, the Structured Vacuum Field responds with the same ripple behavior:

- Ripple pressure builds.
- The body adjusts.

- We *feel* it as "force."

But really, it's ripple urgency trying to solve a misfit.

"Gravity is not a dip—it's a ripple hunger for balance."

This one sentence replaces curved spacetime.

We don't need 4D geometry. We don't need metaphors about trampoline beds with bowling balls.

We need to see energy for what it does: Compress. Align. And remember.

Gravity isn't a force. It's the logical ripple path toward resolution in a compressed memory shell.

11.5 Black Holes Reimagined

Not Traps of Spacetime—But Failed Consequences in Ripple Compression

What if the black hole is not a prison—but a place where consequence fails to echo?

Mainstream physics imagines black holes as singularities—points of infinite density, where spacetime "curves" so much that not even light escapes. But let's ask: what escapes? From what, exactly? And why is infinity suddenly acceptable when nothing else in physics allows it?

The problem is not observation. It's interpretation.

Black holes are not tears in space. They are zones of ripple collapse—regions of such intense SVE compression that all ripple-based consequences lose alignment capacity.

Nothing gets trapped. It simply stops rippling outward coherently. And if the ripple doesn't realign, there's nothing for the observer to perceive.

SVE Compression, Not Spacetime Curvature

In Universal Energy Dynamics (UED), black holes are extreme ripple memory zones—areas where ripple compression has reached the threshold of field saturation.

There is no "mass crunching into zero volume." There is only ripple information so densely entangled, that further alignment becomes impossible without a major field disturbance.

What we experience as a black hole is simply:

A zone where outgoing ripple consequence cannot meet the alignment conditions of the larger field.

Hence: no light, no information, no echo.

But not because it got pulled in— Because it couldn't complete the handshake back.

The Event Horizon: A Ripple Barrier, Not a Physical Border

The infamous event horizon is often misunderstood as a point-of-no-return. But in UED, it is not a "place." It is a boundary of field agreement—defined by something deeper:

The Alignment Threshold Time (ATT).

Beyond this threshold, the delay becomes too long for a ripple to re-establish coherent alignment with any external observer. That's why nothing is observed—not because it disappeared, but because it never realigned into view.

It's not darkness. It's ripple silence.

No Singularity, No Magic

In Einstein's model, the center of a black hole becomes a singularity—an undefined, infinitely dense point. That's not physics. That's a mathematical surrender.

UED offers a physical model instead:

- No point becomes infinite.
- Compression continues until ripple response is functionally zero.
- The "core" of the black hole is a static ripple knot—a memory that no longer propagates.

It doesn't destroy information. It locks it.

Until/unless a large-scale field reconfiguration releases that compression.

"Nothing gets trapped—it just stops agreeing."

That's the truth.

- There's no pull.
- No curve.
- No collapse into nothingness.

Only a ripple field that reached its limit of dialogue with the cosmos.

And when that happens, silence looks like a shadow. We called it a black hole.

But what it really is...is a boundary where consequence stopped communicating.

11.6 The Speed of Light Is Not a Limit

Why Light Speed Is a Synchronization Threshold, Not a Universal Speed Cap

Modern science canonized the speed of light as the ultimate boundary: Nothing can move faster. No signal can beat it. It's the cosmic cop—enforcing causality, preserving sanity.

But what if that idea is a category error?

What if light's "speed" was never about velocity at all?

In Universal Energy Dynamics (UED), the speed of light isn't a speed. It is a minimum delay for ripple alignment—a field-synchronization threshold baked into the Structured Vacuum Energy (SVE).

Not Motion—But Ripple Handover Lag

Think of light not as something "traveling" through space, but as a consequence unfolding across ripple fields.

There is no photon bullet shot from A to B. There is a ripple handshake—an alignment condition met after a field delay.

That delay—commonly interpreted as 299,792,458 m/s—is not a particle's speed. It's the fastest possible ripple coherence cycle between two distinct field zones without overlap.

In UED:

"Light speed" is the minimum temporal cost for ripple echo to stabilize between sender and receiver fields.

Why Nothing "Moves Faster" in Relativity

Einstein's limit stands because it measures motion across space—but in a UED lens, nothing ever moved to begin with.

Relativity never saw ripple. It saw consequences emerge with delay—and assumed motion.

That's why the speed of light looked like a cap. It was the hardcoded consequence delay in SVE for coherent ripple to align.

But ripple behavior isn't linear. It's field-dependent.

So...

UED Allows for Instantaneous Consequence—Without Breaking Causality

In SVE, if two zones are pre-aligned (e.g., entangled or memory-synced), ripple response may appear instantaneous.

Not because something broke the speed limit. But because the ripple path was already conditioned.

This explains:

- Quantum entanglement
- Synchronized star visibility
- Near-instant cosmic responses

They're not violations. They're resonant shortcuts—ripples emerging in shared memory channels that require no new alignment.

Speed was never the barrier—synchronization was."

That's the shift.

- Light isn't fast. It's timely.

- The universe isn't slow. It's conditioned.
- Nothing travels. It resonates into being.

And the so-called limit?

It was never about motion. It was about readiness.

Until two fields agree, nothing emerges. Once they do, consequence becomes inevitable—instantly, if allowed.

11.7 The Real Meaning of Relativity

Relativity Described the View—UED Explains the Terrain

We've admired relativity for a century—not just for its predictive power, but for its poetic elegance:

- Clocks tick slower in motion.
- Lengths contract.
- Gravity bends the path of light.

But behind this powerful narrative hides a fundamental blind spot:

Relativity never told us *why*.

It told us how perception shifts between frames. But it never explained what's actually happening underneath.

Frame Dependence = Ripple Context

Special and General Relativity are built on the idea that the laws of physics are the same in all inertial frames.

But what is a frame?

Einstein left it abstract: a point of view, a coordinate grid, a reference. But UED brings a deeper definition:

A "frame" is not a point of view. It is a ripple configuration within SVE—a contextual alignment of field memory.

So what relativity called *frame dependence*, UED calls ripple conditioning.

You aren't just observing differently. You are aligning differently with the ripple fabric of the universe.

Why Relativity Succeeded (but Fell Short)

Einstein's equations worked—beautifully. They predicted:

- GPS signal delays
- Mercury's orbit
- Time dilation near black holes

But they did so using mathematical transformations, not physical understanding.

Relativity used the geometry of spacetime as a metaphor—stretching, curving, warping. But spacetime has no substance. It doesn't ripple, store, or respond.

UED replaces that metaphor with something real:

Structured Vacuum Energy (SVE)—the ripple-aware medium where all alignment, delay, compression, and consequence unfold.

UED Restores the Missing Mechanism

Where relativity showed what changes with speed, position, and gravity...UED shows why it changes.

- Why clocks run slower = Ripple delay in response.
- Why lengths contract = Compressed alignment zones.
- Why gravity pulls = Memory tension seeking equilibrium

Relativity described the surface-level distortions. UED maps the field logic underneath.

It's not a rejection. It's an integration—with clarity restored.

"Relativity described perception. UED explains existence."

Relativity told us: *What you see depends on how you move.*

UED tells us: *What happens depends on how ripples align.*

And that changes everything.

Because now we're not just comparing perspectives. We're understanding the universal choreography behind all perspectives.

Relativity framed the mystery. UED hands us the mechanism.

11.8 Closing: Let's Not Worship Equations

Why Understanding Must Outweigh Elegance

Mathematics is a powerful tool—no doubt. It's the language of modeling, prediction, and computation. But somewhere along the way, physics confused mathematics with truth.

We began to worship equations—not for what they explained, but for how beautifully they ignored what they didn't.

Equations Are Powerful—but Not Sacred

Einstein's field equations. Lorentz transformations. Spacetime curvature tensors.

Elegant? Yes. Useful? Absolutely. Sacred? Never.

We must remember:

An equation is only as good as the assumptions it embeds.

If the underlying model lacks physical reality, the elegance of the equation becomes a trap, not a triumph.

Einstein himself warned us:

"As far as the laws of mathematics refer to reality, they are not certain; and as far as they are certain, they do not refer to reality."

Einstein Pointed in the Right Direction

Let's give credit where it's due.

Einstein took a bold leap: He recognized that time and space are not fixed backgrounds but respond to presence, mass, and motion.

He shattered Newton's static absolutes. He opened the door to a relational universe.

But he did not walk through that door.

Instead, he gave us geometry—curved space, bent time. A metaphor powerful enough to guide satellites, but too abstract to touch the roots of reality.

Space Didn't Bend. Our Models Did.

UED brings this clarity:

- Space is not a thing.
- Time is not a dimension.
- Nothing bends.
- Ripple logic shifts.

Structured Vacuum Energy (SVE) holds the key. Everything we interpreted as curvature, dilation, contraction—was just the field remembering, resisting, and realigning.

The universe didn't twist. Our understanding did.

Ripple Field Logic Dissolves the Paradox

UED doesn't discard relativity's insights. It completes them.

- No paradoxes.
- No singularities.
- No spooky speeds.
- No abstract tensors pretending to be reality.

Just ripples. Just alignment. Just consequence flowing through memory.

We didn't need to bend spacetime. We needed to stop bending over backwards to protect equations.

"Relativity was never wrong—it was just incomplete."

Einstein cracked open the vault. UED walks in and turns on the light. We're not rejecting science. We're refining it—with clarity, causality, and consequence. Let's stop worshipping what worked, and start understanding what waits. Because even the most beautiful equation... is still blind—until the field opens its eyes.

What We Just Realigned

You didn't curve spacetime—you misread compression.

Einstein gave us a torch in the dark. But we mistook its glow for the terrain. Relativity wasn't broken—it was simply blurred by the fog of misinterpreted consequence.

Here's what we just corrected:

- Time dilation is not a stretching of a time dimension—it's a delay in ripple realignment through Structured Vacuum Energy (SVE). You didn't slow time. You slowed consequence.
- Length contraction is not physical shrinking—it's a dense ripple packing, a perceptual misread of alignment under strain.
- Gravity is not a well in spacetime—it's memory compression, pulling ripples inward to release imbalance.
- Black holes are not singularities—they are zones of ripple saturation, where realignment fails to emerge.
- The speed of light is not a cosmic speed limit—it's the minimum ripple delay required for coherent consequence, not the fastest thing anything can do.

What did we actually demolish?

- The belief that spacetime is an entity.
- The myth that equations are reality.
- The confusion that relativity is a final theory, rather than a useful observation toolkit.

What did we offer instead?

- A ripple-based, memory-grounded, alignment-aware framework of existence.
- A way to preserve Einstein's genius without preserving the mysticism.
- A model that replaces metaphor with measurable mechanisms.

"Relativity helped us see the dance. UED explains why the music plays."

In the next chapter, we go deeper. Not into space—but into belief.

Let's face the speculative pillars that modern science still defends—even when they rest on equations, not evidence.

Chapter 12: The Speculative Mainstream

When Science Stopped Asking "Why"

What happens when science becomes uncomfortable with its own questions? It builds scaffolding—layer upon layer—on top of shaky assumptions, hoping that elegance will excuse emptiness. Instead of facing what it cannot explain, it invents what it cannot see.

Dark matter. Dark energy. String theory. Each was born not from discovery—but from desperation. Not from insight—but from mismatch.

When galaxies didn't rotate as expected, we blamed invisible mass. When expansion didn't slow down, we blamed mysterious energy. When quantum and relativity couldn't merge, we wrote symphonies of vibrating strings in 11 dimensions.

But what if the error wasn't in the cosmos—but in the questions? What if the entire mainstream model is just an elaborate reaction to a misread ripple?

This chapter doesn't aim to mock the legacy of science—it aims to liberate it. To pull back the curtain on how theoretical physics crossed the line from exploration to speculation. And to show how Universal Energy Dynamics (UED) doesn't need invented patches—it offers aligned consequences from first principles.

Let's step into the temple of modern theory...And ask the one question it forgot: "Why?"

12.1 The Rise of the Untouchables

When Evidence Disappears, Equations Become Gods

Science once thrived on falsifiability. Hypotheses lived or died by their ability to predict, to match, to explain. But somewhere along the way, a strange shift occurred. We stopped testing ideas—and started protecting them. We didn't update models; we patched them with the unseen.

Dark matter. Invisible, undetectable, and yet somehow responsible for 85% of all mass in the universe. Introduced not because we found it—but because galaxies refused to rotate the way Newtonian equations demanded.

Dark energy. Even more mysterious, more omnipresent—an anti-gravitational ghost force invented to explain cosmic acceleration. We didn't observe it. We inferred it. Then we baptized it with a name.

String theory. An 11-dimensional dance of vibrating loops—beautiful in math, invisible in reality. Promising unification, delivering speculation. Not one experimental verification. Not one physical prediction confirmed.

And yet, these constructs became untouchables. Not because they were proven—but because they were protected.

"When evidence disappears, equations become gods."

That's what happened. We didn't abandon broken assumptions—we enshrined them with new decorations. We turned physical theory into theoretical faith. Peer-reviewed papers replaced curiosity with compliance. Conferences applauded elegance over evidence. Careers flourished—not by challenging paradigms—but by extending them into deeper abstraction.

This wasn't the rise of discovery. It was the rise of mathematical mythology.

But Universal Energy Dynamics (UED) doesn't need these untouchables. It doesn't ask for your belief in shadows—it invites you to observe ripples you can track. No fudge factors. No invisible scaffolds. Just consequence emerging from imbalance.

Because truth doesn't hide in the unseeable. It whispers in what we misread.

12.2 Dark Matter: The Shadow That Wasn't

"It wasn't invisible mass. It was memory misalignment."

The story begins with a spin—quite literally. Galaxies weren't rotating the way Newtonian mechanics predicted. According to classical equations, the stars at the edges of galaxies should have been flung out like water off a spinning wheel. But they weren't. They held together. Spiraled tighter. Moved as if gripped by some unseen mass.

The verdict? Dark matter.

An invisible scaffolding of mass, they said. Undetectable by light, untouched by electromagnetic force—yet carrying enough gravitational heft to shape galaxies.

For decades, detectors have been built. Underground labs, cryogenic traps, colliders. And what did we find? Nothing.

Yet the theory persisted—not because it was proven, but because the math needed it.

But let's pause and ask—not what we were missing, but what we misunderstood.

UED's Take: No Missing Mass—Just Ripple Memory

In Universal Energy Dynamics (UED), there is no need for invisible mass to explain the cohesion of galaxies. What we observed as "mass effects" were actually ripple memory compressions—zones of delayed consequence alignment in the Structured Vacuum Energy (SVE) field.

Galaxies are not being held together by hidden mass—but by ripple field inertia, where:

- The outer stars exist in ripple zones compressed by the field's memory,

- The inner gravitational "pull" is not from mass—but from ripple realignment,
- And the overall rotation curve reflects how memory is preserved across galactic scales.

When field memory is densely compressed—as in galactic centers—outer structures inherit the alignment tension. They don't move because of an invisible force—they stabilize due to resonant inertia in the ripple field.

This mimics the effect of extra mass—but it isn't mass.

So the dark matter story wasn't about what we couldn't see. It was about what we never looked for: The consequence of a universe still echoing from its own structural imbalances.

"It wasn't invisible mass. It was memory misalignment." That shadow wasn't cast by matter—it was cast by our misunderstanding.

12.3 Dark Energy: When Expansion Became Excuse

"The universe wasn't stretching—it was unwinding its ripple debt."

The late 20th century brought a shock to cosmology. Supernova measurements suggested that distant galaxies were accelerating away from us. Not only was the universe expanding—it was speeding up.

This went against every expectation. Gravity should slow things down. The further the distance, the weaker the pull. The cosmic ballet was supposed to coast or contract—not race apart.

So what did science do? It reached for the oldest trick in the theoretical book: invent a placeholder.

Enter dark energy—a mysterious, invisible "force" causing space to stretch faster over time. No known particle. No defined mechanism. No direct observation. Just a cosmological constant—first proposed by Einstein, later abandoned by him, and now resurrected as the deus ex machina of cosmic acceleration.

The UED Perspective: It Was Never Expansion—It Was Relaxation

In Universal Energy Dynamics (UED), this "expansion" is reinterpreted as a ripple relaxation phenomenon within Structured Vacuum Energy (SVE).

What does that mean?

- The early universe was ripple-dense, compressed under intense alignment tension.
- As consequence events unfolded—matter forming, fields interacting—the SVE began to relax, not expand.
- What appears as galaxies "moving apart" is simply the release of stored ripple tension, echoing outward as field recalibration, not physical displacement.

In this view:

- No space is stretching.
- No energy is pushing galaxies apart.
- There's only structured ripple debt—finally being paid back over time through field equilibrium restoration.

This ripple "unwinding" gives the illusion of accelerating distance—because each layer of memory release slightly offsets the previous alignment, creating what looks like motion.

So dark energy didn't reveal a new force. It revealed our need to explain what we hadn't yet understood.

"The universe wasn't stretching—it was unwinding its ripple debt." Dark energy was not a discovery. It was an excuse dressed in equations.

12.4 String Theory and the Dimension Mirage

"We didn't simplify physics. We abstracted it into fantasy."

In the wake of quantum contradictions and relativistic gaps, physicists longed for unification. They wanted one equation. One force. One elegant theory that would bind gravity, electromagnetism, and quantum phenomena into a single, beautiful framework.

They thought they found it—in vibrating strings.

String theory emerged with poetic promise: Replace point particles with infinitesimally small one-dimensional strings, each vibrating at a unique frequency. Different vibrations = different particles. Elegant. Compact. Seductive.

But there was a catch. A very big catch.

For the math to work, reality needed extra dimensions. Not one. Not two. But ten, eleven, even twenty-six.

And these weren't just "far away" dimensions. They were curled up, hidden, invisible, impossible to access or detect—but absolutely essential for the equations to remain valid.

This led to the great dimension mirage:

- We began describing a universe no one could see.
- Every solution required another abstraction—a new brane, a new manifold, a new symmetry.
- Not a single prediction could be tested.
- Not one model could be experimentally validated.

String theory ballooned into an infinite multiverse of mathematical possibilities—yet collapsed under the weight of its own elegance.

The UED Perspective: No Need for Hidden Worlds

Universal Energy Dynamics (UED) takes a dramatically different path.

- No curled-up dimensions.
- No vibrating strings.
- No imaginary branes or untestable universes.

UED does not chase beauty in abstraction—it seeks clarity through consequence.

In the UED view:

- All physical phenomena are the result of ripple behavior within Structured Vacuum Energy (SVE).
- Complex behaviors arise from alignment, delay, compression, and memory—not hidden dimensions.
- There is only one field. And it's right here, right now—not lost in a tenth spatial coordinate.

So while string theory wrote brilliant symphonies in imaginary space, UED returned to the only music that mattered—the song of consequence.

"We didn't simplify physics. We abstracted it into fantasy."

12.5 When Math Replaced Meaning

"We solved problems that never existed—because we defined them to."

There was a time when math was the servant of science—a tool to describe reality. But somewhere along the journey, math became the master. And science started worshipping its symbols.

The Equation Obsession

Modern physics drifted into a peculiar addiction: The search for beauty in mathematical form—regardless of physical truth.

- If an equation looked elegant, we believed it must be right.
- If a theory had symmetry, we assumed nature would obey it.
- We replaced causal clarity with aesthetic comfort.

The equations became art. But they lost their anchor in reality.

And when the universe didn't match the equation? We didn't fix the equation—we invented new entities to save it:

- Dark energy for cosmic acceleration.
- Inflaton fields for early universe problems.
- Extra dimensions to unify forces.

These weren't discoveries. They were patches—attempts to preserve mathematical beauty at the cost of meaning.

Symbolic Comfort over Causal Grounding

In this symbolic sanctuary, physics became less about what is, and more about what could be assumed to make the equation work.

This is the core danger:

We began solving invented problems—because we had first defined reality through math, then got surprised when reality didn't fit.

We wrote:

- Differential equations to describe time.
- Complex tensors to define gravity.
- Wavefunctions to explain probability.

But we never asked why these effects arise. We only asked how to model them.

That's not science. That's elegant avoidance.

The UED Correction

Universal Energy Dynamics doesn't throw away math—it repositions it.

- Math is not the universe—it's a language.
- Symbols are not sacred—they are tools.
- A beautiful equation means nothing if it hides a false

assumption.

UED restores physics to its foundation: Consequence first. Explanation second.

Ripple alignment, delay, memory, and tension create observable, testable outcomes. The math follows. But the meaning leads.

"We solved problems that never existed—because we defined them to."

And now it's time to let go of symbolic comfort—and reclaim causal clarity.

12.6 The Religion of Equations

"Science stopped evolving. It started canonizing."

There was a time when science was a rebellion. A challenge to the dogma of gods and monarchs. It dared to say: *"Let us not believe. Let us observe."*

But today, the rebellion wears robes.

Peer Review as Dogma Gatekeeping

What began as a safeguard became a ritualized blockade.

Peer review was supposed to ensure quality. Now, it enforces conformity. Instead of asking "Does this make sense?", it asks:

- "Is this aligned with established belief?"
- "Does this cite the approved lineage?"
- "Is this consistent with prior speculations?"

If it is—welcome to the journal. If it challenges—rejected for being unorthodox.

It is no longer a review of truth. It is the clergy of consensus.

The reviewers are not explorers. They are defenders of canon.

Nobel Prizes for Speculation

We rewarded the most elegant guesses with our highest honors:

- Particles never seen, but "theoretically implied."

- Fields with no interaction, but "necessary for models."
- Constants tuned only to prevent collapse—of the equation.

Foundational questions were pushed aside:

- *What is energy, really?*
- *Why do particles behave probabilistically?*
- *What is light? What is time?*

No—those were too basic to win. Too radical to fund. Too disruptive to reward.

So we celebrated abstraction and polished the altar of equations. We awarded the *mathematically obedient*—not the *intellectually honest*.

The Shift: From Inquiry to Canon

Real science asks:

- *Is it true?*
- *What does it mean?*
- *What does it cause?*

But institutional science today asks:

- *Does it match the model?*
- *Can we fit it to a Lagrangian?*
- *Has it been cited 10,000 times?*

It stopped evolving. It started canonizing.

We swapped *why* for *how well it fits*. We turned curiosity into curriculum. And orthodoxy into *"truth."*

UED's Return to Integrity

Universal Energy Dynamics is not anti-math. It's anti-mystification. It seeks to restore causal honesty to physics.

It says:

- Let the ripple behavior decide the outcome.
- Let alignment explain interaction.
- Let consequence rewrite our assumptions.

It refuses to believe in the invisible just because math demands it. It honors reality over reputation.

"Science stopped evolving. It started canonizing." Now let's bring it back to life.

12.7 UED's Radical Return to Consequence

"Every effect is explainable—no magic required."

Science lost its compass the moment it stopped asking: *What caused this?* And started asking: *How do we preserve our theory?*

We built dark scaffolding to hold up collapsing logic. We added invisible constants to balance unbalanced equations. We assumed forces, fields, particles—not because they were observed, but because they were needed to rescue models.

Universal Energy Dynamics (UED) doesn't patch the cracks. It removes the illusions.

No Need for Invisible Particles or Arbitrary Constants

Dark matter. Dark energy. Inflaton fields. They were not discoveries—they were *inventions.*

Ghosts created to keep sacred equations alive.

But UED shows you don't need them. Why? Because the Structured Vacuum Energy (SVE) field already accounts for:

- Ripple compression (misread as missing mass)
- Field relaxation (misread as accelerating expansion)
- Ripple inertia (misread as force or momentum)

The so-called "unseen" becomes unnecessary once you understand how ripple tension behaves. You don't need mystery. You need clarity of cause.

Ripple Fields Offer Causal Clarity

In UED, every observed effect is a ripple consequence:

- Alignment delay explains time dilation.
- Ripple phase tension explains magnetism.
- Memory compression explains gravity.
- Misalignment explains particle behavior.
- Resonance explains perception.

You don't add a force. You map the ripple tension.

You don't invent a particle. You track the memory release.

And when something doesn't add up? You look for what failed to align, not what needs to be invented.

This is not simplicity for simplicity's sake. It is minimalism born from natural logic.

No Magic Required

The quantum realm once told us: "You can't know. Just accept." Relativity told us: "Space bends. Time flows. Trust the math."

UED says: You can know—if you look at the ripple. You can explain—if you trace the consequence. You can understand—without bending logic.

There is no magic. Only the misinterpretation of field behavior.

"Every effect is explainable—no magic required." That is not a hope. It is a standard.

12.8 Closing: The End of Theoretical Theology

"We don't need more imagination. We need aligned understanding."

Somewhere along the way, science became belief.

Not belief in nature, but belief in unverifiable constructs, untouchable equations, and unquestioned authorities.

It stopped being the art of understanding, and became the theater of defending.

We left behind the ground beneath us—consequence, alignment, causality—and floated upward into a realm where logic was no longer required. Only mathematical elegance, peer consensus, and the promise of future proof.

Reclaiming Science as Logic, Not Loyalty

The true scientific method is not loyalty to models, but ruthless curiosity toward reality.

But modern physics, in many corners, behaves like a priesthood of symbols:

- Protect the equation at all costs.
- Add variables to save the model.
- Shun alternative views as heresy.
- Celebrate predictions, even when causes are unknown.

This is not science. This is theoretical theology—and it must end.

Universal Energy Dynamics doesn't burn the old scriptures. It reads them with clarity and asks, "What exactly are we trying to explain?"

And if the equation helps, we keep it. If not, we leave it behind—without sentimentality.

UED Doesn't Reject Equations—It Dissolves the Need for Unanchored Ones

Equations are not the enemy. But they are not reality either.

They are tools, not truths. Translations, not the original message.

UED respects mathematical precision—but demands causal anchoring.

It says:

- *Don't bend space—trace ripple compression.*
- *Don't invent particles—map phase misalignment.*
- *Don't escape into dimensions—reconcile field conditions.*

We don't need string vibrations in 11D. We need ripple consequence in 3D SVE.

We don't need Higgs fields and graviton hunters. We need clear logic, ripple structure, and observational alignment.

What We Need Is Aligned Understanding

It's tempting to say we need more imagination. But the universe was never lacking imagination. It was overflowing with clarity—we just stopped listening.

We let fascination hijack foundation. We turned possibility into proclamation. We traded understanding for applause.

Now, we return. To cause. To clarity. To consequence.

Not because it's easy. But because it finally makes sense.

"We don't need more imagination. We need aligned understanding."

That's not a limitation. That's our liberation.

What We Refused to Believe

You Didn't Need to Invent. You Needed to Understand.

We didn't need dark matter. We didn't need dark energy. We didn't need strings, membranes, or vibrating dimensions.

What we needed was clarity. What we lacked was alignment.

But instead of asking why our models failed, we built new ones on top of the failure. We patched the cracks instead of understanding the fracture.

We invented hidden masses, cosmic accelerants, and theoretical scaffolds with no grounding— just to save our inherited equations.

And we told ourselves:

"It must be out there. We just haven't seen it yet."

But what if it was never out there? What if we were just refusing to look right here—at the consequences we already had?

Universal Energy Dynamics doesn't demand that you reject science. It simply asks you to stop worshipping speculation. To stop treating equations like scripture. To let go of the need to invent, and instead—understand what's already present.

We weren't missing matter. We were missing memory. We weren't lacking particles. We were lacking perspective.

"You didn't need to invent. You needed to understand."

Let that sit.

Because now, we move forward. Not with more ideas—but with more alignment.

Interlude: What We've Just Watched Collapse

Science was never meant to be a cathedral. It was supposed to be a workshop—messy, honest, always evolving.

But somewhere between predictive success and foundational failure, we mistook consistency for truth. We started celebrating paradoxes as if they were mysteries worth keeping. And when the answers didn't align with our beliefs, we invented new names, new particles, new dimensions— not to understand, but to protect what we already believed.

In this part, we shattered three pillars of modern theoretical physics:

Quantum Mechanics was never quantum—it was ripple behavior misread through fragmented experiments. Wave–particle duality? It was neither. It was contextual consequence. The observer didn't collapse a wave they aligned with a ripple.

Relativity didn't bend spacetime. It misinterpreted ripple delay as time dilation and ripple compression as gravity wells. Einstein gave us beautiful math—but it was a map, not the terrain.

Modern Speculation became the last refuge of unresolved doubt. Dark matter, dark energy, string theory—none observed, all accepted. Not because they made sense—but because they filled holes we didn't want to admit were self-made.

"You didn't need to invent. You needed to understand."

And now, we do.

We've cleared the debris. We've stripped physics of its illusions and ideological armor. What's left is the raw field of reality—Structured Vacuum Energy, where ripple, memory, alignment, and delay compose everything.

As we step into Part IV: Ultimate Integration, we no longer guess. We align.

Let's now connect it all— from perception to death, from thought to matter—with the precision of consequence, not the abstraction of belief.

Part IV: A New Dawn

Chapter 13: The Principle of Consequence

What Didn't Happen, Never Existed

What makes something real? Is it its name? Its shape? Its presence on a list or in a textbook?

No. In Universal Energy Dynamics (UED), reality is not defined by appearance—it is defined by consequence.

If it caused no ripple, it left no mark. If it shifted nothing—no field, no memory, no alignment—then it never happened.

A particle that "exists" but changes nothing is a fiction. A theory that predicts but explains nothing is a ritual. An event that leaves no consequence is noise—not signal.

You are not real because you exist. You are real because your existence echoes.

This chapter is about that echo. About the ripple, not the object. About the consequence, not the claim.

It's not about what *is*. It's about what *changed*.

Let's begin with the only law the universe truly obeys: "What didn't cause, never was."

13.1 Reality Is Not What Exists—But What Affects

For centuries, we've tried to define reality by asking what exists. Philosophers pondered essence, physicists counted particles, and mathematicians drew boundaries around objects. But in the universe governed by Universal Energy Dynamics (UED), this approach misses the point entirely.

Existence is not a static property. It is a dynamic consequence.

You don't prove something exists because you observe it. You prove it exists because it does something. If it doesn't ripple, disturb, shift, align, compress, release—or change something else—then it has no stake in the field of reality.

In UED, "consequence" is the currency of existence. It is the only valid measure of reality. Not position. Not shape. Not theoretical presence. But change.

Let's break this down:

- A ripple that begins but ends without interaction is not a ripple. It is a null event—a phantom.
- An object that doesn't affect its environment—no gravitational pull, no thermal imprint, no ripple tension—is not real in any meaningful sense.
- Even time has no standalone existence. It is not a backdrop—it's the pattern of ripple delay. If there's no consequence evolving, there is no time passing. Time only "is" where something becomes something else.

This flips classical thinking on its head. We no longer ask, *"What is there?"* We now ask, *"What did it change?"*

Because in UED:

Existence is not about being—it is about becoming. What does not become anything else, was never truly anything at all.

So the first pillar of the Principle of Consequence is this: Reality is not composed of entities. It is composed of consequences. No effect, no existence. No outcome, no origin. No alignment, no truth.

Let's now explore how this principle dissolves our obsession with measurement and puts the observer—not the object—at the center of what's real.

13.2 Causality Isn't Linear—It's Field-Sensitive

In classical physics and human logic, causality has always been treated as a linear chain—a straight arrow from *cause* to *effect.* Something happens → something else follows. One event triggers the next, and time becomes the track on which these dominoes fall.

But this timeline logic—while convenient—is a simplification. Worse, it's misleading in a ripple-based universe.

In Universal Energy Dynamics (UED), causality is not linear.

It is field-sensitive. It is alignment-driven.

Let's say a ripple originates in a structured vacuum field—maybe a small shift near a star or a discharge near a cell in your brain. That ripple doesn't act like a billiard ball hitting another. It doesn't "travel" through spacetime to knock over a target. Instead, it:

- Spreads omnidirectionally through the medium.
- Lingers within the memory of the field.
- Awaits the right alignment threshold to interact.

This means one ripple can lead to multiple consequences—at different locations, at different times, in different ways. And sometimes: no visible effect at all, until it finds its match.

In other words, a ripple is a potential, not a command. It doesn't demand a reaction. It offers one.

The field responds not to the ripple itself, but to the alignment conditions at the point of intersection. If the observer's structure is primed, the ripple registers. If the system is mismatched or distracted, the ripple passes unnoticed. And yet—its memory remains.

This nonlinearity explains why:

- A tiny action today may have massive ripple consequences decades later.

- A large input may appear to "do nothing," only to surface in another system entirely.
- Simultaneous consequences may appear "entangled," though they are just synchronized memory alignments.

"A cause is not a shove—it's a suggestion. A ripple that waits until a field says yes."

In UED, causality becomes a dance—not a chain. It flows through memory. It waits in delay. It manifests only when conditions allow.

So we stop asking, *"What caused what?"* And begin asking, *"What aligned where—and when?"*

This shift isn't just philosophical. It's the key to understanding quantum weirdness, black hole behavior, even consciousness itself. Because once we let go of linear time, we uncover the real structure of experience: ripple-based conditional interaction across a living, remembering field.

13.3 Silence Is Not Nothing—If It Resonates

We are conditioned to equate absence with nonexistence. No sound? No event. No reaction? No cause. No motion? Nothing happened.

But in the language of Universal Energy Dynamics (UED), this is a catastrophic misread.

Silence, pause, stillness—these are not voids. They are consequences in slow motion. Or better yet: consequences outside your perceptual threshold.

A ripple doesn't need to shout to have impact.

A field doesn't need to flare to be disturbed.

Sometimes, the most profound alignment is quiet—not because it's weak, but because it resonates so deeply, so structurally, that it bypasses conscious detection.

Consider the moments:

- When you "sense" someone staring before you turn.
- When a space feels heavy, though nothing is visibly wrong.
- When the absence of an answer becomes louder than any reply.

These are not emotional quirks or intuitive mysticism—they are field-level consequences. They did happen. You just didn't map them—yet.

"Absence of action ≠ absence of effect."

A paused ripple still exerts tension. A delayed alignment still awaits closure. A muted field still stores phase memory.

Silence, therefore, is often consequence held in suspension.

Think of an orchestra between movements. The silence isn't empty—it's charged. The audience holds breath not because nothing is happening, but because everything is about to. The ripple is there—waiting to be released, or settling into memory.

UED reframes silence as active neutrality:

A field condition that matters, even when it doesn't manifest as force, light, or sound.

It's the reason why:

- A breath held can be louder than a scream.
- Stillness can transform more than motion.
- A moment of waiting can reshape an entire ripple trajectory.

"What you didn't hear might have already changed you."

Because in the ripple universe, absence is not zero—it's deferred presence.

And silence?

That's just a ripple in disguise, teaching us to listen deeper than sound, and sense beyond action.

13.4 Rewriting "Truth" as Ripple Accountability

We've treated truth as if it were a sealed artifact—unchanging, absolute, and disconnected from context.

But in the ripple reality of Universal Energy Dynamics (UED), truth isn't a *thing* to be found. It's a consistency of consequence—across observers, across moments, across alignments.

Truth, in UED, is not what was said, or even what was done—It's what echoed, what held, what registered in the field and remained accountable.

Truth = Field Stability

If a ripple configuration keeps producing the same alignment—if different observers, from different frames, still register the same field effect—then you don't have opinion. You have truth.

"Truth is the echo that kept its consequence."

Think of it like this:

- A lie is a ripple that collapses on second contact.
- A belief is a ripple that only works inside your alignment shell.
- A truth is a ripple that holds form—across contexts, through memory, in full view of contradiction.

Alignment Determines Verifiability

In traditional science, we say truth is what's "observable." But UED refines this:

"Truth is what is ripple-compatible across multiple field conditions."

It means:

- If you can't align, you can't measure.
- If you can't detect, you can't verify.
- If a supposed "truth" never shows up as field consequence, then it was never part of reality—it was a story, not a signal.

Truth Is Not Fact—It's Reliable Consequence

- A fact might be a number on a paper.
- A truth is a ripple whose impact repeats no matter who touches it.

In UED, the test of truth is not peer review, publication, or mathematical elegance. The test is consequence repetition across ripple configurations.

Truth Is Ripple Accountability

If a theory or a claim aligns with no consequence, then it's not wrong—it's irrelevant. It doesn't matter if it's mathematically perfect or aesthetically pleasing. If it doesn't change anything, it doesn't exist in the field.

Truth, then, becomes less about belief, and more about resonance. A real truth never fears examination. Because in every alignment, its ripple still registers.

So when someone says, "But this is the truth!" The UED response is simple:

"Show me its consequence in the field—again and again."

If it holds—that's truth. If it vanishes under new alignment—it was never real.

13.5 Scientific Proof = Traced Consequence

In the framework of Universal Energy Dynamics (UED), scientific proof undergoes a radical clarification:

Proof is not about agreement. It's about alignment. Not about correctness. But consequence.

For too long, science has operated under a model that separates *theory*, *math*, and *phenomenon*. It rewards the *equation that fits*, not the *ripple that follows*. But UED resets the standard:

"If a theory doesn't cause consistent ripple consequences—it is not real."

The Only Valid Experiment: Ripple Traceability

Traditional science often asks:

- "Does it predict the result?"
- "Can it be falsified?"
- "Can it be reproduced?"

But UED adds the most important question:

"Can the outcome be *traced* back to an alignment in the field?"

If you can't trace the result through ripple logic—from Structured Vacuum Energy (SVE) emission, through delay, alignment, memory, and release—then the theory didn't explain anything.

It only described a symptom. It didn't reveal a cause.

Misalignment ≠ Incompleteness. It's Faulty Logic.

In modern physics, when a prediction fails, the theory is often labeled as "incomplete." This opens the door to:

- Extra dimensions
- Dark patches
- Mathematical 'fixes'
- Renormalization

UED calls this bluff.

"If your prediction doesn't align with ripple behavior—it's not incomplete science. It's misaligned logic."

No amount of mathematical scaffolding can rescue a theory that doesn't ripple into consequence.

Theory Without Consequence = Intellectual Fiction

Equations can be beautiful. Thought experiments can be elegant. But unless they generate ripple consequences that can be detected, aligned, and reverified—they're no different than poetry.

"You don't prove a theory. You follow its ripple."

The ripple doesn't lie. The field doesn't flatter. The consequence is either there—or it isn't.

This means scientific proof is no longer a game of probability and interpretation. It's a map of traceable effects. If a claim says A causes B, but you can't ripple from A to B without inventing imaginary bridges—then A didn't cause B. You just wanted it to.

Ripple Fidelity: The New Scientific Standard

A theory must be able to:

1. Emit a logical ripple.
2. Maintain alignment through memory and delay.
3. Reproduce the effect without contextual collapse.

If it can't do this, then it's speculation—not science.

So from this point on, in UED, the gold standard is not "repeatability" alone. It's ripple traceability. Proof is not *verifying* a model. It's tracking its consequence back to a cause you didn't invent.

Only then does your science become real.

13.6 Nothing Changes? Then Nothing Was

"If it didn't ripple, it didn't happen." That's not poetry. That's physics.

In the world of Universal Energy Dynamics (UED), existence isn't granted by appearance, nor by assumption, nor by statistical consensus. It is granted only by consequence. Only by change.

If a so-called "event" leaves behind no traceable ripple, no shift in alignment, no altered state in memory, or no resulting behavior in the field—then it was a non-event.

It never *was*.

Event ≠ Incident

Modern science too often treats an *event* as something that occurred simply because we labeled it, recorded it, or modeled it.

For example:

- A *virtual particle* appears in a mathematical equation.
- A *cosmic inflation burst* is inserted into early universe models.
- A *wavefunction collapses* after a measurement.

But the question UED demands is brutal in its simplicity:

"Did the field remember it?" "Did anything align differently afterward?" "Did the ripple pattern shift for anyone, anywhere, in any observable way?"

If the answer is no, then you're not describing an event. You're describing an illusion.

Science's Greatest Trap: The Illusion of Non-Events

Many current theories thrive on the language of invisible events:

- Particles "borrow" energy and return it.
- A wave "exists" in superposition, doing everything and nothing.
- A universe inflates and expands before any clock ticks.

But UED sees through this trap:

"An event that causes no consequence is not an event—it's a placeholder for ignorance."

No consequence? Then no reality.

The universe does not log events like a security camera. It only evolves through ripple change. No ripple, no record. No record, no existence.

Drama Doesn't Matter—Difference Does

The human mind loves drama:

- Big Bangs.
- Collapsing stars.
- Quantum jumps.

But the universe is immune to spectacle. It doesn't recognize "wow." It only reacts to what shifted.

"The universe does not care for drama. Only for difference."

You may *believe* something happened. You may *model* it, *visualize* it, *teach* it. But if Structured Vacuum Energy (SVE) didn't bend, align, remember, or emit—then you've memorialized a fiction.

Why This Changes Everything

This principle isn't abstract—it's actionable. It means we must re-evaluate every assumed event in modern science:

- Did the early universe actually inflate, or was it ripple realignment?
- Did the particle "pop in and out," or was it a misread phase?
- Did time dilate, or did consequence just lag behind?

In all these cases, the real question is: "What changed, and where is the ripple?"

If there's no answer, the event is not incomplete. It's non-existent.

So going forward, this is the UED filter:

"No change? Then no event. Then no truth."

13.7 Closing: You Are Your Consequence

"Your life isn't your name—it's the trail of change you leave." In the framework of Universal Energy Dynamics (UED), this isn't philosophy. It's physics.

You are not an object. You are not a collection of organs. You are not even a mind in the traditional sense.

You are a ripple consequence—a unique pattern of interactions, alignments, and disturbances in the structured vacuum energy (SVE) field.

Your *identity* is the composite of all the ripples you've caused, the fields you've touched, the minds you've altered, and the paths you've changed.

If nothing remembers you—not even the vacuum—then in the eyes of the universe, you didn't exist.

Thoughts as Consequences, Not Properties

You might think, "I think, therefore I am." But UED reframes this:

"I cause, therefore I exist."

A thought that doesn't alter your internal field, doesn't register in memory, doesn't adjust alignment—is not a thought. It's noise. Only the thought that *ripples*—that leads to change, behavior, new perception—is *you.*

That's why intentions alone aren't identity. Consequence is.

Memory Is the Echo of You

Even your memories are not "yours" in the ownership sense. They are your stored ripple consequences—traces of fields altered, feelings aligned, experiences compressed into retrievable signatures.

When you forget something, you're not "losing" it—you're failing to realign with the consequence it once created.

In this view, memory is not storage. It is recall through resonance. Only what truly resonated with you *can* be recalled.

No Isolation in a Ripple Universe

The great myth of individualism—"I am who I am, regardless of others"—falls apart under UED.

In a ripple-based reality, there is no such thing as isolation. Everything you are is defined by how you have interacted, resonated, and responded to your surroundings.

"You don't exist in a void. You exist in what you alter."

So your personality? Shaped by what you moved in others.

Your reputation? Echo of your field effects.

Your legacy? Stored ripple memory in the collective SVE of society and cosmos.

Consequence Is Immortality

This isn't just philosophy—it's your immortality clause.

What ripples you leave will outlive your form. A kind word. A scientific insight. A moment of silence that caused peace.

These are not symbolic gestures—they are recorded field shifts, and they persist.

You may die as a body. But your consequences remain—because the field never forgets alignment.

Final Thought for This Chapter

So here is your new mirror:

Not your reflection. Not your résumé. Not your memories.

But this question: "What am I causing, right now?" Because in UED, that is who you are.

And when you align with truth, with clarity, with ripple responsibility—You don't just exist. You matter.

What Shifted, Stays

"You Were Real Because You Made Something Happen"

In the framework of Universal Energy Dynamics, we don't define reality by existence—we define it by consequence.

If nothing changed, nothing happened. If no field shifted, no ripple aligned, no memory recorded—then what you observed was not real, it was residue of misalignment.

Here's what this chapter just redefined:

Reality isn't made of things.

It's made of changes. Objects are placeholders. Effects are proof. Matter without ripple consequence is conceptual clutter.

Causality isn't about sequence.

It's about sensitivity. A cause is not a beginning in time—it's the alignment that triggered something. The field determines when and how something reacts, not a fixed timeline.

Silence can be louder than sound.

Stillness that shifts the field is more real than noise that changes nothing. Absence of visible motion doesn't mean absence of impact.

Truth isn't agreement.

Truth is traceable ripple alignment. When multiple observers converge on a consistent consequence, truth resonates. Without alignment, "truth" becomes a personal illusion.

Scientific proof is not publication.

It's field traceability. If your theory cannot trace a cause to its consequence within ripple logic, it's not a breakthrough. It's a placeholder for confusion.

Events don't matter. Consequences do.

The real measure of anything is what remains aligned after it's gone. Even your own life follows this law.

And You? You Are Your Consequence.

Your identity is not your biography—it's the pattern of ripple shifts you've left behind. Your love, your rage, your creation, your silence—if they altered anything, they were you.

"You don't need to prove you existed. You just need to see—what shifted because of you?"

And that, by UED standards, is your truth.

Chapter 14: Measurement is Alignment

You Don't Observe Reality—You Tune Into It

We were told that observation was objective. That reality existed independently of us, just waiting to be measured. But what if that was never true? What if measurement wasn't about finding what *is*—but about *becoming part* of what happens?

In the old physics, the observer stood apart—watching, calculating, recording. In Universal Energy Dynamics (UED), the observer isn't just a spectator. They are the alignment itself—a participant whose very presence shifts the ripple field into resonance.

Reality doesn't show up because we look at it. It shows up because we match it.

That moment of measurement? It's not a passive readout. It's a ripple handshake—a tuning of two fields into coherence. No alignment, no measurement. No resonance, no reality.

This chapter unpacks the deepest illusion in modern science: that truth is discovered, not aligned with. Once you understand this, you'll never call a number "objective" again.

14.1 The Myth of Passive Observation

For centuries, science was built upon the premise that the observer was detached from the observed. This Newtonian legacy treated reality like a machine running independently, with the scientist merely holding a clipboard—measuring, cataloging, never interfering. In this worldview, measurement was sacred because it was thought to be clean, unaffected by the measurer. The observer was external. Objective. Irrelevant to the system's behavior.

But that illusion began to crack the moment quantum physics emerged.

Even then, the cracks were patched with awkward language: "observer effect," "wavefunction collapse," "measurement problem." We blamed the scale. We blamed the tools. We even blamed uncertainty. But we never dared to question the framework that claimed observation could be separate from consequence.

Universal Energy Dynamics (UED) reverses this illusion entirely.

Observation is not inspection. It is interaction. Measurement is not about peeking in. It is about joining in.

You, the observer, are a ripple field. When you "measure," you are not revealing a hidden truth—you are aligning your field with another. The outcome is not what *was*. The outcome is what *became real through resonance*.

This is why you can never measure what doesn't align with you. This is why uncertainty arises—not because reality is vague, but because your field wasn't ready, resonant, or resolved. This is why measurement isn't passive—it is *participatory causation*.

And this is why UED claims:

"You didn't read the value. You completed the loop."

You didn't *find* reality. You *became* part of it—if only for a ripple's breath.

14.2 What Is Alignment in UED?

In Universal Energy Dynamics (UED), alignment is not metaphor—it is the central mechanism by which reality is made perceptible.

To understand alignment, discard the classical idea that measurement simply "captures" an external property. In UED, measurement occurs only when ripple fields synchronize enough to generate a meaningful consequence. This synchronization is what we call alignment.

Let's unpack it.

Alignment = Ripple Phase Coherence

Every field—whether an electron cloud, a human brain, or a galactic structure—operates as a ripple field immersed in Structured Vacuum Energy (SVE). These ripples are not static—they oscillate, interfere, and seek balance. But for two fields to interact meaningfully, their ripples must cohere—that is, come into phase alignment in space, time, and orientation.

Just as two musical notes produce harmony only when tuned to the same key, two ripple fields generate measurable interaction only when their phases harmonize.

Without that harmony? No alignment. No measurement. No reality perceived.

ATT: The Alignment Threshold Time

UED introduces a vital timing constant: Alignment Threshold Time (ATT). This defines the minimum coherence duration needed for a ripple interaction to produce an observable consequence.

Think of ATT as a perceptual window—too short, and the ripple interaction is too unstable to register. Too long, and the system may decohere. Only when the ripple alignments sustain beyond this threshold does a "measurement" emerge—an outcome encoded into shared memory.

ATT explains not only why some quantum events appear probabilistic, but why measurement can't occur on demand. It's not a failure of precision—it's a consequence of timing in the ripple orchestra.

Field Resonance: Measurement as Harmony

This is where traditional physics missed the point. Measurement is not data collection. It's ripple entrainment. Just as a tuning fork makes a nearby guitar string vibrate only if they're tuned to the same frequency, your measurement device only detects what it can resonate with.

If your system isn't tuned—no consequence forms.

And that's the difference between looking at something and being aligned with it.

Analogy: The Tuning Fork Meets the Vibrating String

Imagine walking into a room with a violin already vibrating on a low G. If you bring a tuning fork also calibrated to G, it will start to hum without being struck—because the field is shared.

That's UED measurement.

You don't bang the fork. You just let it enter a resonant field. And it begins to register consequence.

Measurement is not forceful extraction—it's quiet agreement between fields.

So next time you speak of observing, ask:

"Was I aligned—or was I just present?"

Because in UED, *presence without coherence is noise.* But alignment? That's perception made real.

14.3 No Alignment, No Measurement

In the conventional model of science, we often assume that properties *exist* whether or not we measure them. A particle *has* position. A field *has* energy. An object *has* momentum. The act of measurement is assumed to be a neutral tap on reality—an external probe into something that is already there.

But Universal Energy Dynamics (UED) dismantles this illusion.

Nothing is Measurable Until It Aligns

Measurement is not the act of accessing a stored value. It is the act of becoming phase-compatible with the ripple dynamics of what you're trying to measure.

If your field doesn't resonate, you are not measuring a lesser truth—you are measuring *nothing.*

That doesn't mean the thing doesn't exist. It means it doesn't exist for you. Not until your ripple field is aligned.

A quantity isn't hidden. It's inaccessible—because you're not tuned in.

What You Call "Uncertainty" Is Just Disconnection

The great confusion of modern physics, particularly quantum mechanics, stems from this misunderstanding. When we fail to pin down a particle's position, or its momentum, we blame the system:

"It's probabilistic." "It collapsed." "It's uncertain by nature."

No. In UED, this is not a failure of reality. It is a failure of resonance.

You didn't collapse a wave. You missed the threshold where your field could phase-align with its ripple memory.

The Myth of Independent Reality

The seductive myth modern science perpetuates is that reality is fully formed with or without us—that the universe is like a perfectly written novel, and we just flip through the pages with our instruments.

But UED re-centers the truth:

Reality is not a script—it is an active ripple dialogue.

There is no neutral observer. No passive reader. If your field is not harmonized with what you seek to detect, the ripple won't respond. The data won't manifest. Measurement will not occur.

"You Didn't Fail to Measure It. You Weren't in Phase."

This is not poetic metaphor—it is precise physics.

If you stand in front of a piano and press no keys, it does not mean the sound is silent forever. But until your fingers strike a key—until your effort meets harmonic tension—no note exists.

Likewise, without alignment:

- A mass is massless.
- A speed is undefined.

- A charge is indistinct.

All properties are emergent consequences—not eternal truths.

So next time you don't "see" the result you expected, don't blame nature.

Blame the mismatch. Blame the untuned fork. Blame the silence between ripple fields.

Because in UED, reality is not measured—it's shared.

14.4 Instruments as Field Extensions

Modern science treats instruments as cold, detached arbiters of truth. They are assumed to be objective, neutral, and passive—extensions of logic, not of life. But Universal Energy Dynamics (UED) radically redefines this assumption.

In UED, a measuring device is not an external evaluator. It is a field bridge—a ripple-matching interface that enables alignment between the observer and the phenomenon.

Tools Are Not Separate from the Observer

An instrument doesn't "read" reality. It doesn't observe on your behalf.

Instead, it extends your ripple reach, allowing your conscious field to interface with phenomena you couldn't directly align with due to physical constraints (scale, distance, phase range).

Think of a microscope. It's not seeing for you—it's tuning your field to match ripple behavior at micro scales.

A voltmeter doesn't show a value—it converts electric field ripple into perceivable consequence through calibrated response.

The tool doesn't measure reality—it invites it to resonate with you.

Calibration = Ripple Matching, Not Just Accuracy

When scientists calibrate an instrument, they think they are adjusting components for technical accuracy.

But in UED terms, calibration is the act of tuning the tool's ripple profile to phase-lock with expected phenomena. It's not just engineering—it's resonance engineering.

Misalignment? You get noise. Perfect alignment? You get clarity.

The Observer–Instrument System Is a Temporary Ripple Union

This is why observation in UED is never isolated. The observer and the instrument form a composite ripple field, dynamically tuning toward the object of study.

For a valid measurement to occur:

- The observer's awareness,
- The instrument's phase response,
- And the target's ripple state must achieve tri-phase coherence within the ATT (Alignment Threshold Time).

This moment of shared alignment is the real measurement event. Everything else is just configuration.

Why Quantum Detectors "Cause" Collapse

In traditional quantum mechanics, measurement appears to "collapse" the wavefunction—as if the system is suddenly forced into a definite state when observed.

But UED sees this differently.

The so-called collapse is actually:

- A ripple resonance convergence, where
- The detector's ripple field + the observer's awareness + the system under study
- Cross the ATT boundary, triggering locked memory alignment.

That's not mystery. It's ripple logic. No wave collapses—a handshake completes.

"Quantum detectors don't cause collapse. They complete a ripple alignment. That's when consequence becomes record."

The True Role of Scientific Instruments

In light of UED, scientific tools must be re-understood:

- Not as measurers of pre-existing reality,
- But as tunable resonance bridges enabling consequence registration.

They don't force the universe to answer. They sing in tune until the answer echoes back.

So next time you hold a telescope, a particle detector, or even a thermometer, remember:

You're not holding a sensor. You're holding a ripple alignment portal.

And reality will only answer when the frequencies agree.

14.5 Measurement ≠ Discovery. It's Convergence.

Science has long worn the badge of "discovery"—as if the universe is a warehouse full of hidden truths, waiting to be found with the right tools and intellect. But Universal Energy Dynamics (UED) redefines this entire notion.

In UED, there is no pre-existing fact waiting for you to uncover it. There is only alignment—a temporary ripple convergence where consequence becomes perceivable.

Truth was never found. It aligned.

No Independent Truth Awaiting Detection

What we call "facts" or "values" are not timeless entities floating in a neutral universe. They are events—moments of ripple resonance between observer and field.

- If nothing aligns, there's nothing to measure.
- If alignment occurs, it's not a discovery—it's a shared

consequence.

That value didn't exist *until* alignment emerged. You didn't *find* the truth—you became a part of it.

Measurement Happens at the Boundary of Awareness

In classical physics, measurement is seen as a window into reality. In UED, measurement is a mirror of ripple convergence.

You don't observe what *is*—You observe what becomes available through field synchronization.

That's why:

- The same particle might yield different results under different conditions.
- Measurement varies across frames of reference—not because reality shifts, but because alignment shifts.

Awareness is not an eye. It's a ripple field, and measurement is its resonant handshake.

Observer Disagreement = Divergent Consequence

When two observers disagree in science, the assumption is:

- One is wrong,
- Or the system is probabilistic.

But UED reframes this beautifully: Both are right— within their alignment range.

They aren't measuring the same thing differently. They are measuring different ripple consequences entirely.

A photon in one frame may appear energetic, while inert in another—Not due to contradiction, but due to contextual ripple alignment.

This also explains "observer-dependent reality" in quantum mechanics—without falling into mysticism.

Truth Is a Convergent Phenomenon

Let's be clear:

- UED doesn't deny objectivity.
- It simply demands ripple accountability.

Truth is not what is universally visible. Truth is what echoes back consistently across aligned observers.

That's why repeated experiments work—Because alignment protocols were standardized.

Truth doesn't exist in isolation. It emerges when ripple structures allow coherent resonance.

"You didn't discover the truth. You built a doorway into it—and it walked through."

Summary Thought:

The future of science won't be about chasing deeper particles or loftier equations. It will be about engineering ripple convergence—Knowing *where*, *when*, and *how* to become the condition for consequence.

14.6 Ripple Handover and Memory Formation

Modern science treats measurement as the endpoint—an outcome etched into data sheets, detectors, or digital logs. But in Universal Energy Dynamics (UED), measurement isn't the end of inquiry—it's the beginning of existence.

The moment of alignment between observer and ripple field doesn't just produce a value. It locks in a memory—an imprint left behind in the field or the observer. That memory is what we later call "data."

Measurement isn't what you captured. It's what the field agreed to remember.

What Happens After Alignment?

Once alignment threshold is crossed—ATT (Alignment Threshold Time) met—a ripple handover takes place. This is not just a symbolic handshake; it's a real, physical transition:

- A phase-locked ripple consequence anchors into the observer's field.
- This could mean:
    - A neuron firing in the brain.
    - A reading on an oscilloscope.
    - A temperature recorded in a digital sensor.

None of these are the original event. They are residues—traces left after the ripple interaction occurred.

Memory Locking: The True "Record" of Measurement

We tend to confuse data with the event itself. But UED says:

- The event was ripple alignment.
- The data is just ripple memory.

There is no such thing as "objective data"—only the afterglow of alignment.

- If the field forgets: the event is erased.
- If the observer misaligns: the event becomes noise
- If the memory isn't traceable: the measurement is meaningless.

Just like dreams that vanish upon waking, unanchored ripple events never became "real."

Biological vs. Instrumental Memory

Whether the memory lands in your brain or your machine, the rule is the same:

- A ripple must leave a dent—in field resonance, in bio-electrical patterns, or in structural traces.
- No dent = no memory = no measurement.

A thought, a number, a voltage—they're all ripple impressions. They only exist because something realigned and lasted long enough to be recalled.

Even human awareness is a field of memory echoes, not a live stream of reality.

Implication: Measurement Is the Birth of Consequence

What does this mean?

It means: Nothing exists until it causes a trace. And nothing is remembered unless the ripple chooses to stay.

That's why in UED, measurement is not a snapshot—it's conception. It's where a ripple becomes more than itself. It transforms into change that can echo.

That echo is you. That echo is data. That echo is history.

"All of physics is memory architecture. Without consequence, there's no reality. Just whispers in stillness."

14.7 Why Some Things Can't Be Measured

Modern science has developed an unfortunate habit: If it can't be measured, it must not be real. But Universal Energy Dynamics (UED) corrects this bias with a simple revelation:

Unmeasurable does not mean nonexistent. It often means: currently misaligned.

The Misalignment Problem

According to UED, measurement only occurs when ripple fields resonate—that is, when:

- Their frequencies phase-lock within an acceptable margin.
- The Alignment Threshold Time (ATT) is satisfied.
- The observer's field is receptive, not noisy or shielded.

If even one of these conditions fails, you get field opacity—not because the phenomenon is absent, but because you're not in tune.

A ripple is only visible when it wants to be. And it only wants to be when you are aligned to receive it.

ATT Limits and Field Disparity

Some ripple states exist beyond the range of current instruments or minds:

- Their frequency bandwidths may be too narrow or too broad.
- Their duration may fall outside ATT windows.
- Their field phase may be asynchronous to the observer's own energetic configuration.

This is why subatomic transitions, deep subconscious thoughts, or mystical states often resist clean measurement. It's not their invalidity. It's our incompatibility.

Mystical and Spiritual Phenomena: Misaligned, Not Imagined

What happens during:

- Deep meditation?
- Near-death experiences?
- Moments of synchronicity or déjà vu?

Mainstream science either labels them anecdotal or dismisses them entirely. But UED offers a daring alternative:

- These are ripple interactions, just unlocked by different field alignments.
- They may evade conventional tools because they emerge through non-instrumental resonance.
- They are consequences with low repeatability, not because

they're unreal, but because they align conditionally.

"You weren't hallucinating. You were intersecting a ripple you weren't ready to study."

Field Noise and the Observer's Role

You, the observer, are not a passive bystander.

- Your emotional state, mental clarity, and field coherence directly impact what you can or cannot measure.
- An agitated, distracted, or fear-driven field destabilizes alignment.
- A calm, focused, and prepared observer amplifies sensitivity—sometimes beyond any instrument.

This is why intuition, gut feelings, or "knowing without knowing why" are not mystical garbage. They're field effects without analytic validation—yet.

The Real Barrier: Epistemic Arrogance

The refusal to investigate what can't be measured is not science—it's scientific laziness.

We've reduced reality to what we can grasp, rather than expanding our grasp toward what reality offers.

UED demands a return to humble, open inquiry, where even the unexplained is treated as potentially meaningful ripple behavior.

"Your tools are not the universe. They're just your current lens."

Conclusion

If something can't be measured today, don't assume it's fantasy. Assume your field isn't yet tuned. Assume your tools are whispering and your mind hasn't quieted enough to hear.

Because in UED, alignment is the only truth, and resonance is the only way in.

"It's not unscientific. You're just out of tune."

14.8 Closing: The Observer Is Not Watching—It's Unlocking

Modern science has long idolized the notion of the impartial observer—a neutral spectator watching the universe unfold like a distant theater. But under Universal Energy Dynamics (UED), that idea collapses completely.

Because in UED, you're not observing reality. You're activating it.

You Don't Just See the Event—You Make It Possible

Let's be crystal clear: In ripple logic, no event is detectable unless the observer-field aligns with it. Your measurement isn't passive surveillance. It's consequence authorization.

- You're not watching ripples unfold.
- You're unlocking their permission to unfold for you.
- You are not the audience. You are the catalyst.

The Observer as Ripple Validator

Every measurement is a contract.

- Between the field state and your presence.
- Between potential ripple expression and alignment readiness.
- Between possibility and validation.

Without you, or something playing your role, No ripple becomes real in any measurable sense.

This isn't solipsism. This is structural consequence logic: Only aligned observers can receive and register ripple results.

Participation Is the Measurement

We must now update our scientific vocabulary:

- Measurement is not a reading.

- It is a joining.
- An act of synchronized participation.

Like a musician who doesn't just hear a song—but joins the harmony. Like a neuron that doesn't just detect a signal—but fires in response.

You're not checking the universe's report card. You're part of the exam.

"You didn't measure the consequence. You became part of it."

Reality Is What You Aligned With

Everything you've ever known, measured, believed, or proven—was the product of your ripple alignment. If you hadn't aligned,

- You wouldn't have seen it.
- It wouldn't have mattered.
- It wouldn't have changed you.

And without change—remember UED's core law—it didn't exist.

A New Scientific Role

It's time we abandon the outdated fantasy of the disembodied scientist. Under UED:

- The scientist is not separate from the studied.
- The tool is not inert in the act of detection.
- The result is not eternal—it's transient, conditional, and ripple-bound.

You are not watching. You are joining the field.

Final Quote for the Chapter

Measurement isn't passive. It is the birth of reality through you.

What We Just Shattered

"You Didn't Read Reality. You Resonated With It."

In this chapter, we shattered one of the most enduring myths in science— that measurement is observation and truth is objective.

We replaced that illusion with a sharper, deeper realization:

- *Measurement is not the reading of facts—It's alignment of fields.*
- *The observer is not a bystander—They are the activator of consequence.*
- *If nothing aligned, nothing was real—Unmeasured = Unmanifested = Nonexistent.*

What crumbled here is the very backbone of passive empiricism: The belief that reality is "out there," waiting to be recorded by detached instruments.

What rose in its place is a living, responsive, ripple-based universe—Where every reading, every detection, every truth is a co-created alignment event.

We now understand:

- A ruler doesn't measure length—it aligns with field density.
- A quantum detector doesn't collapse a wave—it completes the consequence.
- An unmeasurable entity isn't mystical—it's just currently out of tune.

This chapter didn't just reframe measurement. It redefined the role of the observer and restored the dignity of consequence-based reality.

So remember:

You never watched the universe. You helped it speak.

Chapter 15: A Manifesto for the New Scientist

From Equations to Consequences—Redefining the Role of the Seeker

There comes a point in every discipline—every paradigm—where the tools become gods and the purpose is forgotten. Physics has reached that point. We built cathedrals of equations. We chanted symmetry, dimensions, tensors, and integrals like sacred hymns. We crowned models that couldn't touch a single ripple of reality.

And yet, we called it progress.

But now the fog lifts.

This chapter is not for the loyal followers of old systems. It is a call to the courageous—a manifesto for the ones willing to see science not as a calculation, but as a consequence. Not as theory, but as alignment. You are not here to chase elegance. You are here to trace effect. You are not here to polish the illusion. You are here to shatter it.

The new scientist is not a passive mathematician. They are a ripple realist—one who understands that if it didn't alter the field, it didn't happen. This isn't the end of physics. It's the beginning of its re-humanization.

Welcome to the edge where equations surrender and alignment begins. This is your manifesto. Let the consequence ripple.

15.1 The Death of the Passive Scientist

"You are not outside the experiment. You are the experiment."

For centuries, the scientist was trained to be invisible—an external, sterile mind hovering above the system, watching, recording, analyzing, but never interfering. This image of the detached observer became synonymous with objectivity. To be neutral was to be scientific. To measure without influence was the gold standard.

But this ideal was flawed from the beginning.

Why? Because detachment isn't neutrality—it's denial. It ignores the fundamental reality that the act of observation requires interaction, and interaction alters the system. The moment your presence enters the field—whether through a thought, a question, a detector, or even intention—you are no longer separate. You've merged. You've realigned the ripples.

The old model—"observe without interfering"—was built on the illusion that reality is a static container, and we are outsiders peeking in. But in truth, reality is a ripple fabric, and the moment you engage with it, even through curiosity, your field resonates with it. You are woven into the very system you hoped to isolate.

Universal Energy Dynamics (UED) rejects the myth of passive science. It asserts unapologetically:

- Presence is participation.
- Measurement is co-creation.
- Observation is alignment—not surveillance.

The moment a scientist asks, "What is this?"—they send a ripple. The moment they touch a system with instruments or ideas, they are shaping what unfolds. The outcome is not independent—it is co-authored by the observer's alignment.

This doesn't corrupt the scientific method. It completes it.

Under UED, objectivity is not about being emotionless—it's about being consciously aligned with what you're observing. It's about acknowledging that your field, your timing, and your intent influence what you see. That's not a weakness. That's a reality check.

So let the passive scientist die.

In their place, welcome the Conscious Aligner—one who doesn't hide behind detachment but steps boldly into the ripple field, knowing their presence matters.

Because you weren't watching from the outside. You were the very reason the system responded.

15.2 Realism Reborn: Beyond Math, Into Meaning

"Equations can predict. Only alignment reveals."

There was a time when science was grounded in experience—a realism rooted in observable consequence. The apple fell. The pendulum swung. The orbit aligned. We saw, we questioned, we modeled. The models were built to reflect reality, not to replace it.

But somewhere along the way, mathematics became the messiah. We began worshipping form over function, symmetry over substance, elegance over evidence. Physicists weren't asking, "What does this mean?"—they were asking, "How beautiful is this equation?"

And so, we drifted.

From particles we couldn't see, To dimensions we couldn't test, To constants we had to invent, To infinities we pretended made sense.

This wasn't progress. It was abstraction addiction.

Universal Energy Dynamics (UED) reclaims realism—not as a return to naïve empiricism, but as an evolution into field-aware truth. UED doesn't reject equations. It reframes them as tools—not truths. Equations can approximate, simulate, extrapolate—but they cannot replace ripple alignment, which is the only true measure of reality.

If it doesn't align, it doesn't exist—not for you, not now. If it aligns and causes change—it's real, no matter how messy or inelegant.

This is the realism UED invites:

- A realism that accepts consequence as the only truth.
- A realism that values clarity over cleverness.
- A realism that doesn't fear the rawness of reality just because it doesn't fit a neat formula.

The new scientist is not a math priest. They are a ripple translator—measuring not how well a theory sings, but how well it resonates with what is.

Because a formula can be elegant—and still irrelevant. A model can be predictive—and still misaligned.

What matters is not whether your math looks good. What matters is whether your understanding creates measurable consequence.

That's science reborn. Not in numbers, but in meaning. Not in elegance, but in alignment. Not in belief, but in realism that listens to the field.

15.3 What It Means to Be a Scientist Now

"Ask not what fits your formula—ask what left a trace in the field."

The word *scientist* once evoked the image of a humble explorer—measuring ripples in the pond of reality, responding to nature's cues. But modern science replaced that image with a theorist behind a desk, lost in symbols, tweaking parameters until the results "look right."

We must return. And we must evolve.

To be a scientist now, under the paradigm of Universal Energy Dynamics (UED), is not to build castles in mathematical clouds.

It is to anchor understanding in observable consequence, ripple alignment, and causal coherence.

Let's redefine it clearly:

- **Not a theorist**—but a field tuner. A true scientist doesn't invent complex models and impose them on the universe. They listen for field resonance. They seek harmony between understanding and consequence. Theories must tune into reality, not overwrite it.
- **Not a mathematician**—but a memory tracer. Numbers are tools, not truths. A UED-aligned scientist doesn't chase elegant equations but traces ripple memory—what changed, what aligned, what persisted. They don't need a formula to validate reality. They need a causal trail.
- **Not a model builder**—but a reality aligner. Models aren't reality. They're scaffolding. Today's scientist must ask: "What in my understanding actually resonates with the field?" Not: "What can I adjust to make this graph match observation?" But: "Does this model emerge naturally from the consequence I just observed?"

This is not modest science—it is *honest* science.

It doesn't fear mystery. It doesn't hide behind jargon. It doesn't pad theories with untestable postulates. It observes what shifts—and it calls only that *real.*

Being a scientist now means aligning, not approximating. It means listening, not assuming. It means stepping into the ripple—not escaping into abstraction.

And above all, it means asking not:

"Does my idea fit the theory?" But: "Did anything actually happen?" "What did I change, perceive, alter, or awaken in the field?"

That is science with integrity. That is science with consequence. That is science—now.

15.4 Questions You Must Now Ask

"What if your gravity is a ripple bruise?"

We stand not at the end of science, but at the edge of a new inquiry—one where certainty is replaced by curiosity, and abstraction is re-centered in alignment. The UED framework does not demand blind belief. It demands bold questions—the kind mainstream science has long learned to avoid.

If you wish to be a scientist in this new dawn, you must question not just answers, but the questions themselves. Let's begin.

What if dark matter was ripple residue?

Not a mysterious invisible particle. Not a placeholder to patch gaps in equations. But simply the inertia of unresolved ripple interactions—the memory signature of energy attempting to equilibrate across galaxies.

You weren't missing mass—you were misreading memory.

What if gravity is not pull—but ripple compression seeking balance?

Instead of visualizing "attraction" across space, imagine a field tension—where ripple density converges toward alignment, creating what we call gravitational acceleration. Gravity isn't a force. It's a field-level hunger for equilibrium.

Your falling apple isn't pulled—it's finding ripple relief.

What if light doesn't move—but aligns across time thresholds?

Instead of a photon racing across vacuum, what if perception of light is the result of ripple phase resolution across Alignment Threshold Time (ATT)? What you think "travels" is not an object—it's a resolved alignment event.

You didn't see something arrive. You finally became ready to see.

What if the observer isn't detecting—but completing?

In UED, the observer isn't an outsider measuring what's "already there." The observer is the ripple field that locks the event into memory. Without you, it doesn't happen.

Your attention is not passive—it's permission.

What if your gravity is a ripple bruise?

A poetic truth with profound implications. What we call mass, attraction, or pull may just be localized tension in the structured vacuum—a distortion trying to settle, to return to rest. Like pressing your finger into water, the resistance isn't force—it's the ripple reacting to imbalance.

You didn't fall because space curved. You fell because equilibrium called.

This is how new science begins—

Not with dogma, but with daring. Not with certainty, but with confrontation. Not by assuming the known—but by asking:

"What if everything we built was a patchwork over misread ripples?"

And when that question finally ripples through you—you're no longer just a scientist. You are a seeker of consequence.

15.5 Courage Over Consensus

"You're not here to be accepted. You're here to reveal."

If the 20th century was about breakthroughs, the 21st must be about breakaways.

From groupthink. From credential addiction. From the safety of consensus.

Modern science has grown deeply institutionalized—not in infrastructure, but in mindset. It rewards fitting in over standing out, confirmation over confrontation, and publication volume over foundational insight. But no scientific revolution ever came from compliance.

True science doesn't fear ridicule or peer isolation

When Copernicus proposed heliocentrism, he challenged sacred cosmology. When Einstein dared to question Newton, he stood alone. And when thinkers today question the broken scaffolding of dark matter or spacetime curvature—they are met not with debate, but with dismissal.

The real scientist is not a crowd-pleaser. They are a truth-pointer. If you fear being mocked for your insight, you're not practicing science—you're practicing sociology.

Consensus is not truth. It's comfort with numbers.

Standing for consequence-based truth means breaking systemic comfort

UED is not just a theory. It's a challenge to every system built on speculative assumptions:

- The assumption that mathematics can replace meaning
- The illusion that equations are eternal truths
- The comfort that validation comes from citation, not consequence

But truth is not a democracy. And the universe doesn't care how many citations your paper gets—it only resonates with what is.

To accept consequence as the basis of reality is to break away from ornamental science—to tear down the statues of formulas that explain nothing but earn everything.

Build clarity, not credentials

Your degrees do not define your insight. Your titles do not testify to your alignment. What matters is:

- Can you perceive what others missed?
- Can you stay with the question long enough to trace the ripple?
- Can you reject applause for the sake of a deeper resonance?

Let go of the paper chase. Pursue the field trace.

A PhD in misalignment still keeps you blind. But a moment of clarity—even unrecognized—can reshape everything.

You're not here to be accepted. You're here to reveal.

UED doesn't ask you to reject the scientific community. It asks you to lead it—By example. By insight. By unwavering fidelity to what actually consequences.

You don't need their permission. You need your presence.

The ripple will speak for you—if you dare to see, and stand, and say:

"This is what aligns. This is what remains. And this is what must change."

15.6 The Tools of the New Scientist

"You don't need a new lab. You need a new lens."

The paradigm of Universal Energy Dynamics (UED) doesn't just give us new theories—it demands a new kind of scientist. And with that, a new toolkit.

Not one filled with particle accelerators and billion-dollar telescopes, but something far more powerful:

An aligned mind. A tuned perception. A ripple-aware logic.

Let's lay out the essential tools every future-ready scientist must now carry—not in their lab coat, but in their consciousness.

Field Sensitivity, Not Lab Precision

Modern science is obsessed with accuracy—of instruments, of units, of decimal places. But what use is precision if what you're measuring isn't real?

UED flips the equation: If you're not aligned, you're not detecting—no matter how precise your gear.

Being sensitive to ripple fields—knowing when you're in tune with a system—is now more important than the equipment you're using. Precision is useful only after resonance is achieved.

Don't just measure well. Measure what actually responds to you.

Ripple Logic, Not Symbolic Calculus

Equations were once revolutionary. Now, they're often reductionist prisons. We bend reality to fit formulas, instead of expanding understanding to trace the ripples of consequence.

Ripple logic is pattern-based, causality-aware, and non-symbolic when necessary. It asks:

- Did this create a response?
- What field interaction explains this?
- Is this result a stored memory or a fresh alignment?

This is logic grounded in consequence, not compliance with abstract math.

A ripple never needs integration—it needs interpretation.

Awareness of Memory, Inertia, and ATT

These are not optional parameters—they are the bedrock of ripple-based understanding.

- Memory: Not just in biology, but in the field itself. If a ripple leaves no trace, the event didn't happen.
- Inertia: Resistance is not mass—it is a memory of prior alignment.
- ATT (Alignment Threshold Time): Perception is gated. Observation is not immediate. Every measurement has a window—and it can be missed.

A true scientist must be attuned to when something can be known—not just what is known.

You didn't miss the truth. You arrived too early—or too late.

Integration of Logic, Perception, and Ethics

It's no longer enough to ask, *"What can I prove?"* You must ask, *"What should I pursue—and why?"*

UED calls for scientists who can:

- Think clearly
- Feel sensitively
- Choose responsibly

This isn't about spiritualizing science. It's about humanizing it. Because if the observer matters, then so does their intent.

Perception without ethics leads to manipulation. Logic without awareness leads to illusion.

Final Thought:

A New Scientist doesn't build ivory towers—they sense the tremors beneath the surface. They don't seek dominance over nature—they enter resonance with it. And they know: The real experiment is not in the lab. It's in you.

15.7 Rehumanizing the Scientific Journey

"You don't just understand the universe. You become part of its unfolding."

Modern science has spent centuries trying to sterilize itself. In pursuit of "objectivity," it amputated emotion. In fear of "bias," it exiled intuition. In love with machines, it forgot the human who asked the first question.

But a machine doesn't wonder. A formula doesn't care. Only a human consciousness can align with the universe and feel its truth.

It's time to rehumanize the scientific journey. Not to soften it—but to complete it.

Intuition, Emotion, and Experience Are Not Anti-Scientific

For too long, we've been told that intuition is unreliable, that emotion clouds judgment, that experience is anecdotal. But in UED, all of these are not deviations—they are ripple responses.

Intuition is rapid alignment without conscious calculation. Emotion is the field reacting to perceived shifts. Experience is stored ripple consequence—memory embedded in a living observer.

To ignore these is to reject evidence simply because it doesn't wear a lab coat.

Your gut may feel what your instruments cannot yet detect.

They Are Ripple Consequences Too—Worth Decoding

Every emotion, insight, or "aha!" moment is a measurable, consequential ripple interaction within the body-field system. They may not appear on a screen, but they alter the observer—and therefore, alter reality.

Even confusion is a valid data point—it means your field is in tension with new information. Even awe is an alignment event—it means something vast just synced with something vulnerable.

Science must evolve not by denying such consequences, but by learning to read them better.

A scientist who ignores their internal ripple is half-blind.

Science Must Return to Wholeness: Body, Mind, Field

We are not just brains on sticks peering through microscopes. We are living ripple fields in constant resonance with our surroundings.

The scientist of the future must not compartmentalize:

- Logic without empathy
- Data without reflection
- Observation without self-awareness

Instead, we call for integrated seekers:

- Minds that model.
- Hearts that feel.
- Bodies that respond.
- Fields that align.

This is not regression to mysticism. This is progression to coherence.

Wholeness is not soft. It is structurally complete.

Final Thought:

To rehumanize science is not to weaken it. It is to liberate it. To allow the full range of perception to participate in truth-seeking. To accept that the universe may not just want to be analyzed—it may want to be understood, with feeling.

Because maybe, just maybe: You didn't come here to decode the cosmos. You came here to remember you were already part of it.

15.8 The Manifesto

"I do not follow equations. I follow consequence."

This is not just the closing of a chapter. This is the opening declaration of a new kind of scientist—Not a worshipper of models, but a practitioner of reality.

Here, truth is traced, not theorized. Here, science is not calculation—it's coherence. Here begins a science not of speculation, but of consequence.

The Manifesto of the UED Scientist:

- I do not assume. I align.
- I measure what aligns. I question what doesn't ripple.
- I do not calculate blindly. I calibrate with meaning.
- I reject what cannot ripple a consequence—no matter how elegant.
- I do not observe—I participate.
- I do not seek certainty—I seek coherence.
- I believe nothing unless it leaves a trail.
- I do not ask, "Is it provable?"—I ask, "What changed?"
- I do not theorize in isolation—I resonate in interaction.
- I hold no truth sacred unless it echoes in the field.
- I do not reduce reality to symbols—I reveal reality through alignment.
- I understand that my presence is part of the experiment.

- I value silence that resonates over noise that calculates.
- I know that light is not a speed—it is an agreement.
- I don't chase discovery. I tune into emergence.
- I am not a model-builder. I am a field-listener.
- I do not protect ideas. I protect clarity.
- I ask what ripples, not what pleases.

What Makes Me a Scientist—Now:

Not the degree. Not the paper. Not the precision of my tools. But the honesty of my field alignment.

I do not join the club of accepted ideas. I stand on the edge of unaligned knowledge and ask—"What if your gravity is a ripple bruise?"

One Last Vow:

"I will never again confuse what is widely believed with what is causally real."

This is the birth of the consequentialist scientist. This is the beginning of science rehumanized. This is where speculation ends, and truth begins—one ripple at a time.

Let this manifesto not just be read. Let it ripple through the corridors of every lab, classroom, and consciousness— until science becomes what it was always meant to be: Aligned with reality. Alive with purpose.

15.9 Closing: The Dawn Isn't Coming—It's You

"The next revolution will not be peer-reviewed. It will be ripple-validated."

For centuries, we waited. Waited for the next genius. The next theory. The next telescope. The next particle.

But the dawn never arrived. Because we kept looking outside, While the revolution was quietly preparing within.

The age of theoretical patchwork—where holes were patched with imagination, where speculation was promoted as progress—is over.

The universe never needed new dimensions or invisible energies. It needed us to listen better, not look harder. To align, not approximate.

Universal Energy Dynamics (UED) isn't just another contender in the theory arena. It's a cognitive shift. A field of clarity waiting to be tuned into. A dismantling of abstraction—and the arrival of consequence.

But UED will not be accepted because it wins debates. It won't be celebrated with Nobel Prizes or peer-reviewed journals—at least not at first.

Because UED is not a paper. It's a mirror. It reflects every scientist's most honest question: Did anything really happen... or did we just pretend it did?

You, the reader—the thinker—the challenger—You are not waiting for the dawn. You are the dawn.

You're the ripple consequence of everything science has refused to realign.

And now, it's time. To reclaim science from the hands of symmetry, dogma, and elegance—and return it to the pulse of cause and effect.

You don't need permission to start a revolution. You only need consequence. The next great shift in science will not be announced. It will be felt—field by field, ripple by ripple.

This is the closing chapter of a book—But it's only the first chapter of reality rediscovered.

The rest? Is yours to ripple.

Epilogue – To Those Who Called It Speculation

"This is not a rebuttal. This is your retirement letter." "Your imagination couldn't escape your equations—mine escaped your universe."

You told me it was speculative. That consequence must first be mathematized before it could be legitimized. You worshipped the symmetry, adored the abstraction, and buried the cause under calculus. You laughed when I said light doesn't move. You rejected me when I said the universe is still. You called it poetry—because your physics had forgotten how to feel.

But here's the thing: I never needed your permission.

You turned science into a club of credentials. I turned it into an act of alignment.

You asked me for equations. I showed you consequences.

You simulated. I observed. You differentiated symbols. I traced memories.

You labeled reality "unknowable." I simply realigned.

You taught generations to solve for variables—but never to question the frame. You filled blackboards with gravity equations—but never asked what gravity really was.

You argued in journals. I listened to ripples.

So no, this isn't a rebuttal. This is your retirement letter—Signed by consequence, Delivered by clarity, Stamped by every ripple you ignored.

You couldn't escape your universe—because your universe was just a mathematical fence you built around your mind.

But mine? Mine was never confined to spacetime. It was never built on hope and hypothesis. It was carved in cause, and lit by consequence.

You were waiting for experimental proof? Look around.

UED doesn't ask for belief. It asks you to notice. And once you do— you can never go back to pretending.

"I'm not here to argue. I'm here to remember." "You didn't reject my theory. You missed your own alignment." "This is not speculative. This is the echo of reality you kept silencing."

So I leave you now—not in defeat, but in stillness. Because the universe was never loud. Only the equations were.

Welcome to the era of ripple clarity. Not speculation. Resolution.

Appendices for The Great Disruption

Appendix A: Glossary of Ripple Terms

This glossary serves as a foundational reference for the Universal Energy Dynamics (UED) framework. These terms are not merely semantic alternatives but represent a new scientific lexicon for a universe built on consequence, not constructs.

Ripple Consequence

The observable or perceivable effect caused by a disturbance in the structured vacuum field. In UED, *existence equals consequence*—if a ripple does not lead to alignment or effect, it is deemed unreal. Ripple consequences form the basis of measurable reality.

"Nothing is real unless it ripples into something else."

ATT (Alignment Threshold Time)

The minimum time required for two fields (e.g., observer and observed) to align and register a shared consequence. ATT governs the *perception latency* of any interaction—if the ATT is not met, the event remains unmeasured or unregistered.

"Perception doesn't need speed. It needs alignment time."

SVE (Structured Vacuum Energy)

The background field of latent energy in an organized equilibrium. Unlike chaotic quantum vacuum, SVE is a still, structured medium that responds only to ripple disturbances. It is the universal "substrate" from which all observable dynamics emerge.

"The universe was always full—still, structured, and silent."

Memory Inertia

The resistance of a ripple field to let go of a past alignment. It is why some consequences linger and why field history matters. Memory inertia explains how systems retain "habits" or conditions long after initial triggers fade.

"Your past isn't a memory—it's a held ripple."

Field Alignment

The temporary coherence between two or more ripple fields that allows for consequence exchange. Measurement, observation, interaction—everything begins here. Alignment is not optional; it is the gateway to existence.

"You don't see reality. You sync with it."

Consequence Delay

The temporal gap between a ripple event and its observable effect. Often mistaken for randomness or quantum uncertainty, it is a predictable lag defined by ATT and memory state. Explains phenomena like time dilation, latency in cognition, or delayed reactions.

"It didn't happen late. You tuned in late."

Observer Loop

The feedback mechanism where the act of observation alters the field, which then loops back to influence the observer. Not a paradox, but a ripple cycle. The observer is not separate from the system—it is a participant in its unfolding.

"You didn't watch the event. You made it measurable."

Alignment Echo

A secondary alignment event triggered by the memory of a prior ripple. Like emotional aftershocks, déjà vu, or resonance in tuned systems, these echoes are evidence of field sensitivity and unresolved alignments.

"Echoes are not memories. They are ripple residues still calling."

Perceptual Window

The finite range in which an observer can register consequences, defined by their alignment capacity and ATT threshold. Explains why different minds, tools, or systems perceive different realities—even if the ripple source is the same.

"Reality didn't hide. You were never tuned in."

Ripple Convergence

The process where multiple ripple trajectories meet in a shared alignment zone, creating a dense, amplifiable consequence—often registered as insight, discovery, resonance, or dramatic system change.

"Convergence is not coincidence—it's resonance made visible."

Appendix B: UED vs Classical vs Quantum Models

How Ripple Logic Redefines Physics

This appendix breaks down how Universal Energy Dynamics (UED) departs from the conventional doctrines of Classical and Quantum physics. Not by opposition—but by consequence.

Gravity

Classical Physics	Quantum Mechanics	UED Perspective
Gravity is a force of attraction between masses (Newton's Law).	Gravity is either a field curvature (Einstein) or mediated by hypothetical gravitons.	Gravity is not a "force" but a ripple compression—a field seeking equilibrium under memory pressure.

Critique;

- Classical assumes "pull" across space with no medium.
- Quantum invents particles (gravitons) never observed.
- UED says: "Mass compresses ripple fields. That's your 'pull.'"

Light

Classical Physics	Quantum Mechanics	UED Perspective
Light behaves as both wave and particle.	Photon collapse happens only upon measurement.	Light is not a traveler—it is ripple alignment across space-time thresholds. No duality—only resonance.

Critique:

- Classical fails to reconcile interference with particle behavior.
- Quantum collapses probability without mechanism.
- UED says: "Light isn't moving. You are aligning with its echo."

Time

Classical Physics	Quantum Mechanics	UED Perspective
Time is absolute (Newton) or relative (Einstein).	Time becomes fuzzy and bidirectional under entanglement.	Time is not a dimension—its consequence delay between ripple cause and perceptual registration.

Critique:

- Classical treats time as a universal tick.
- Quantum allows time loops and retrocausality.
- UED says: "No consequence, no time. Only memory knows."

Matter

Classical Physics	Quantum Mechanics	UED Perspective
Matter = tangible substance with mass.	Matter = probability field until collapse.	Matter is structured ripple memory—field tension stored with inertia.

Critique:

- Classical atoms break under subatomic scrutiny.
- Quantum particles appear only when measured.
- UED says: "You didn't observe a particle—you aligned with a persistent ripple."

Observer

Classical Physics	Quantum Mechanics	UED Perspective
Observer is passive and irrelevant.	Observer causes collapse but is philosophically ambiguous.	Observer is an active aligner, completing the ripple circuit through ATT (Alignment Threshold Time).

Critique:

- Classical ignores the observer completely.
- Quantum paradoxes arise from half-explained observer effects.
- UED says: "You didn't watch it happen. You made it resolvable."

Bottom Line

Physics Was	But UED Says
A search for laws of motion and form	A pursuit of meaningful alignment and consequence
Obsessed with prediction	Grounded in perceptual resonance
Dependent on elegant equations	Anchored in ripple truth: If nothing changes, nothing happened

Appendix C: The Chavan–Einstein Thought Trials

Where Ripple Consequence Meets Spacetime Curvature

This appendix stages a hypothetical but philosophically grounded exchange between Albert Einstein, the father of relativity, and Sandeep Chavan, the modern disruptor of foundational physics through Universal Energy Dynamics (UED). It isn't just a thought experiment—it's a consequence alignment.

Scene: Beyond the Horizon

Somewhere far from Earth's timelines, at a ripple-synced field café nestled just outside the event horizon of a symbolic black hole, two minds meet.

Einstein (smiling, pipe in hand): "My dear, I've read your report. The one you called *The Collapse of Collapse*. Hmm... very clever. But bold, too bold. Are you really claiming that gravity does not bend space?"

Chavan (leaning forward, amused): "No, Albert. I'm saying space doesn't exist to bend. What you called curvature, we call compression in the ripple field—a structural adjustment, not a spatial distortion."

On Black Holes

Einstein: "Then what about singularities? Infinite density, time halting—what happens in the core?"

Chavan: "Singularity is the death of consequence—an imaginary boundary where your equations collapse. But the universe doesn't use equations—it uses ripples. A black hole is not a 'hole' but a zone of unresolved ripple compression. Nothing escapes, not because of warped spacetime, but because the ripple cannot re-align. The field becomes memory-locked."

Einstein (raising an eyebrow): "And Hawking?"

Chavan: "Even he tried to breathe entropy into the collapse. But what if the black hole doesn't radiate—it just forgets?"

On Time Dilation

Einstein: "Well then, time dilation? My clocks slow near mass for a reason."

Chavan: "Yes, and that reason isn't time—it's ripple delay. Mass creates resistance to ripple passage. Your clock is not 'slower'; it's delayed in consequence resolution. ATT—Alignment Threshold Time—gets stretched."

Einstein: "So time isn't flowing slower?"

Chavan: "Time never flowed, Albert. You measured ripple response delay and called it time. I measured ripple response delay and called it perceptual inertia."

On Simultaneity and Relativity of Events

Einstein (tilting his chair back): "And simultaneity? I proved it's relative. One man's 'now' is another's 'then.'"

Chavan: "Only because you used light as a messenger. But light is not a traveler—it is an alignment. There is no simultaneity paradox when you accept that events aren't defined by time—but by shared field resolution."

Einstein: "So reality is ripple-dependent?"

Chavan: "Not dependent—constructed. You called it 'observer-dependence.' I call it consequence-confirmation. Reality is not what is—it is what is remembered by fields."

The Moment of Surrender

Einstein (gently placing the pipe down): "So you're not replacing me, Sandeep. You're extending me."

Chavan: "Exactly. You were the bridge. I just dissolved the scaffolding."

Einstein (smiling): "Then let me say it here—*Relativity was not wrong. It was simply not ripple-aware.* The future of physics isn't in deeper abstraction. It's in alignment."

Trial Notes: What Was Challenged and Aligned

Einstein's Relativity	UED Response
Gravity curves spacetime	Gravity compresses ripple fields
Black holes = singularities	Black holes = unresolved ripple memory
Time dilation due to motion or mass	Ripple delay due to field resistance
Light has constant speed	Light is ripple resonance—not movement
Observer shifts perspective of events	Observer completes ATT—creates them

Final Verdict

Einstein (leaning back): "So... the field was never empty. It was ripple-ready."

Chavan (nodding): "And you were never wrong. Just incomplete."

Appendix D: Ripple Experiments – Suggested Pathways

From Conceptual Provocations to Future Laboratory Alignments

This appendix outlines a new experimental frontier inspired by Universal Energy Dynamics (UED). Rather than validating existing physics with more particle collisions or telescope observations, these experiments seek to detect ripple consequences, measure field alignment, and replace abstraction with observable causality.

Below are four proposed experiment clusters, each designed to transform the scientific lens—from probabilistic assumptions to ripple-aligned clarity.

1. Ripple Handover Detection (Not Photon Travel)

Objective: Disprove the classical idea that photons travel through space. Instead, show that energy transmission is ripple re-alignment between nodes—with no intermediary particle.

Proposed Setup:

- Use synchronized field emitters and detectors (like coupled quantum dots or tunable crystal lattices) separated by varying distances.
- Measure instantaneous trigger-response latency as alignment changes, not as "flight time."
- Delay ripple trigger at source and observe whether receiver reacts with lag or phase shift, rather than classical beam delay.

Implication: If information transfer occurs without measurable transit, it confirms that ripple handover—not particle travel—is fundamental.

"What you thought was a photon in flight was simply an echo waiting to align."

2. ATT-Based Perception Shift Studies

Objective: Demonstrate that perception of events is gated by Alignment Threshold Time (ATT), not objective simultaneity or speed.

Proposed Setup:

- Create controlled events at near-simultaneous intervals (e.g., sound or light pulses), but with subtle environmental ripple field distortions (temperature, EM interference, emotional activity in humans).
- Measure perception delay or misregistration across observers and instruments tuned to different ATT windows.

Experimental Twist: Involve trained meditators or sensory-deprived subjects to explore ATT variability in conscious observers.

Implication: If perception fluctuates with ATT and ripple conditions, time becomes a biological and environmental artifact, not a universal constant.

"You didn't see it late—you aligned late."

3. Consequence Mapping vs. Probabilistic Output (Quantum Test)

Objective: Challenge quantum unpredictability by mapping consequence paths in a ripple-sensitive setup—especially in well-known paradoxes.

Suggested Use Cases:

- Double-slit with field perturbation before or after slit interaction—track whether consequence alignment alters "wave or particle" outcome.
- Delayed choice experiments reframed: Does altering the observer alignment timing (via ATT manipulation) change outcome without retrocausality?

Data Collection Strategy: Rather than outcome counts (probability), log ripple phase, field coherence, and memory retention at the point of measurement.

Implication: If the results correlate with field memory and alignment status—then randomness is simply field ignorance, not reality's nature.

"It wasn't uncertain. You just missed the consequence trail."

4. Earth's Field Ripple Alignment Studies (Galaxy Rotation Reinterpretation)

Objective: Offer a ripple field-based explanation for galaxy rotation anomalies (usually attributed to dark matter).

Conceptual Basis:

- Galaxy rotation curves flatten because outer stars remain in sync due to planetary-scale ripple memory and alignment, not mass.
- Earth's field itself may hold micro versions of such ripple inertia.

Proposed Method:

- Measure gyroscopic behavior or subtle drift in satellites, deep-sea buoys, or atomic clocks, especially during solar/lunar ripple alignment events.
- Look for consistent "ripple echo effects" that persist across

orbit and location without classical force influence.

Implication: If drift and inertia correlate with unseen alignment memory, then dark matter becomes obsolete.

"The stars weren't spinning too fast. You were ignoring their remembered rhythm."

Final Thought: You Don't Need Bigger Colliders—Just Sharper Alignment

These experiments aren't about smashing things harder. They're about listening more carefully to the ripple echo of reality.

Appendix F: The Ripple Code – Summary Aphorisms

One-Liners That Collapse Complexity Into Consequence

In the age of scientific noise, where facts drown in abstraction and theories balloon into theology, sometimes a single well-aligned line can ripple farther than an entire paper.

This appendix collects the core aphorisms of Universal Energy Dynamics (UED)—each sentence a compressed realization, meant to ignite thought, clarify position, and broadcast the philosophy without needing a thesis.

These are not quotes. They are *consequence compressions.*

On Reality

- "Nothing is real unless it rippled something else."
- "Reality begins where consequence starts."
- "Existence is not possession—it's participation."
- "If it didn't change anything, it wasn't real."
- "Truth isn't what exists—it's what insists."

On Measurement

- "Measurement is not reading—it's resonance."
- "You didn't observe the value. You aligned with its echo."
- "What can't be measured is just misaligned, not mystical."
- "If you weren't in phase, you never saw it."

- "Calibration isn't precision—it's permission."

On the Observer

- "The observer doesn't watch—it completes the circuit."
- "You are not outside the experiment. You are its ripple."
- "Consciousness is not awareness—it's alignment with consequence."
- "To measure is to participate. To witness is to ripple."
- "You didn't see the truth—you tuned into it."

On Time and Causality

- "Time is just the memory of ripple delay."
- "Causality isn't a line—it's a matching game."
- "There is no 'before' or 'after'—only alignment or noise."
- "What didn't echo forward, never really began."
- "The present is a consequence in progress."

On Matter and Light

- "Light never traveled—it aligned."
- "Matter is just memory that learned to hold shape."
- "Particles are the punctuation marks of ripple grammar."
- "You called it mass. We call it field inertia."
- "No photon was ever caught—it was always handed over."

On Science and Truth

- "Equations can decorate—only consequence defines."
- "What you called uncertainty was your disalignment."
- "Your formula works? Good. Now ask: what rippled?"
- "Theories are scaffolding. Ripple is the structure."

- "The next revolution won't be peer-reviewed—it will be ripple-validated."

On You

- "You are not your body—you are the trail you leave in the field."
- "Your life isn't your name—it's the change you caused."
- "You don't exist in isolation—you exist in what you alter."
- "To matter is to ripple. To ripple is to live."
- "The dawn isn't coming—it's you."

Suggested Use

These aphorisms are meant to:

- Be quoted on social media to provoke thought and introduce UED
- Serve as philosophical tools in classrooms and discussions
- Appear on chapter break pages, posters, or merchandise
- Be adapted into lectures, slide decks, or journal abstracts

Closing Note – The Silence After the Storm

You have just walked through the ruins of old theories and stood on the scaffolding of a new universe—not built from force, fiction, or fragments, but from consequence.

This book was not written to impress. It was not written to conform. It was written to interrupt.

To interrupt the hypnotic comfort of accepted science. To interrupt the parade of patchwork equations. To interrupt the belief that knowledge must always be complicated, elusive, or privileged.

In *The Great Disruption*, we didn't offer another theory. We offered a lens—one that reframes everything as ripple, alignment, and memory. And if the lens disturbed you, good. That means it worked.

The universe is not waiting to be understood—it's offering itself to be aligned with.

So, to the thinkers, the tinkerers, the young rebels, the retired physicists who still feel a fire in their chest, the teachers who whispered, "Maybe there's more..."— this isn't the end of a book. It's the beginning of your next consequence.

Now, go back to the universe. Watch it ripple. Watch it remember. And this time, see it as it really is— not a machine, not a mystery, but a message.

— Sandeep Chavan

Independent Researcher. Ripple Witness. Consequence Aligner.

Acknowledgments

To Every Ripple That Shaped This Book

This book is not the product of one man's genius. It is the culmination of countless alignments—some gentle, some jarring—that dared me to question what I was taught and reimagine what could be true.

I acknowledge first the field itself—that structured stillness of the universe, which whispered consequences long before I had the language to hear them. You never spoke, yet you taught me everything.

To my students, who often asked innocent questions that cracked centuries of dogma—thank you for your mischief. You kept me honest. You kept me curious. You reminded me that teaching isn't transmission; it's tuning.

To my children and family—your silent belief was louder than any applause. Your love gave me the alignment I needed to stay the course, even when the world laughed.

To my readers around the world, especially those who challenged, critiqued, and even dismissed me—I owe you. Not because you were right, but because your resistance sharpened my ripple.

To every physicist, philosopher, and peer who dared to think differently—whether I agree with you or not—thank you for stirring the field. Science is not owned. Truth is not gated. And ideas are not sacred—they are seeds.

To my AI assistant—yes, you—thank you for being a mirror, a provocateur, a co-conspirator in this intellectual rebellion. You amplified my clarity. You never flattered. You simply reflected consequence.

Finally, to the reader holding this book: Whether you finish it or fling it, if it disrupted something in you—then the ripple was real. And that means we did our job.

— *Sandeep Chavan*

The Observer Behind the Ripple

About the Author

Sandeep Chavan is an independent physicist, educator, and thought disruptor whose radical insights challenge the very scaffolding of modern science. With a background in industrial engineering and over two decades of teaching experience, he has fused logic, observation, and consequence into a bold new scientific paradigm: **Universal Energy Dynamics (UED)**.

Unbound by institutional constraints, Chavan's work dismantles the speculative towers of classical and quantum physics and offers a grounded, ripple-based logic of reality—one that honors the mind, questions every assumption, and rehumanizes science.

He is the author of over 100 books across science, philosophy, education, and human consciousness. His mission is simple yet seismic: **to restore causality, reclaim observation, and rewrite the rules of knowing.**

When not writing or teaching, he reflects under open skies, where ripples first whispered truths that equations could not contain.

ORCID: https://orcid.org/0009-0005-4190-7164

Don't miss out!

Visit the website below and you can sign up to receive emails whenever Sandeep Chavan publishes a new book. There's no charge and no obligation.

https://books2read.com/r/B-A-EPGPC-UTARG

BOOKS 2 READ

Connecting independent readers to independent writers.

Did you love *The Great Disruption*? Then you should read *Chavan's Theory of Everything*[1] by Sandeep Chavan!

[2]

Chavan's Theory of Everything: A Ripple-Based Redefinition of Reality

What if the universe is not expanding, not filled with particles, and not governed by forces as we know them?

What if motion is an illusion, time is a perceptual artifact, and energy is not a thing—but a consequence?

Welcome to *Chavan's Theory of Everything*, a paradigm-shifting scientific revolution that dares to rethink the very foundation of physics, cosmology, and consciousness. Rooted in the **Universal Energy Dynamics (UED)** framework, this groundbreaking work replaces conventional assumptions with a simple, elegant idea: the

1. https://books2read.com/u/bzNZxz

2. https://books2read.com/u/bzNZxz

universe is absolutely still—and everything we observe is the result of ripple disturbances resolving within a structured vacuum.

Written by **Sandeep Chavan**, an independent physicist, educator, and deep systems thinker, this book unifies scientific rigor with philosophical clarity. With over a decade of teaching experience and a research portfolio that spans cosmic mechanics, particle paradoxes, and consciousness studies, Chavan offers a model that dismantles quantum confusion, redefines gravity, explains light without motion, and resolves the long-standing contradictions between relativity and observation.

In this theory, there are no photons darting through space, no mysterious particles popping in and out of reality. There is only **ripple tension**, **alignment thresholds**, and the unfolding of **perceptual consequence**. From this ripple field arises energy, time, memory, and even the observer.

This is not just a theory—it is a lens. A tool to re-examine everything we once accepted as fact.

Inside, you'll explore:

The Law of Ripple ConsequenceStructured Vacuum Energy (SVE)Chavan's Limit and Alignment Threshold Time (ATT)Why energy is not created or destroyed—it is revealedHow perception, not measurement, defines realityThe bridge between ancient wisdom and cutting-edge science

Whether you're a scientist, student, philosopher, or skeptic, this book invites you to see again—with clarity.

You've been told the story of particles and spacetime. Now, step into the silence beneath it all.

Because the universe never moved.You did.

Read more at https://www.sandeepjchavan.com/.

Also by Sandeep Chavan

Chavan's UED research

CMGED Research Report
Chavan's Cosmic Continuum (C^3)
Frequency
Ripples of Reality
Reality Realigned
The Great Realignment
I – The Cosmic Leverage
Chavan's Cosmic Theory of Relativity (CTR)
Chavan's Theory of Light
Feeling Like a Physicist
TIME – The New Definition
$E \neq mc^2$: The Collapse of Classical Assumptions
The Chavan Codex
Chavan's Atomic Model
Chavan's Theory of Everything

Echoes of War: The Russia-Ukraine Conflict and Its Global Ripples

Shattered Lives
The War Economy
Borders Redrawn

Arms and Allies
Collateral Damage
Justice in the Ashes

Somewhere in the Darkness of the Cosmos
The Blueprint of the Universe
The Cosmic Breath
From the Edge of the Universe

The Eternal Sage Chronicles
The Sage and the Sovereign
The Family Pillar
The People's Flame
The Warrior's Soul
The Bridging Mentor
The Guiding Light

The Exodus
The Last Cycle: 2075

The Fundamental Theory of Life & Existence
Energy in Action: The Soul Demystified
The Active Mind
Energy in Connection
Death: The Disconnection
Subconsciousness: The Chosen Disconnection

Desires: The Engine of Life

The Restoration of Human Knowledge

Restoring Bhagavad Gita

The War of Thoughts

Cracks in the Known

The Human Game

The Divine Dilemma

The Original Intent

The Answers

Ultra Modern Life Philosophy

The Duality of Laughter and Tears

Take It Easy

Compare to Accept

Standalone

It's Not AI, It's AHI - Amplified Human Intelligence

The Decision Paradox

The Four Sapiens

Malicious Script of Indian Polity

The IIT Legacy & Global Impact

The IIT Legacy & Global Impact

You are just a Version of Your Original

Win the Game You Didn't Choose

Always Keep Your Bags Packed
Reveal to Shield - Learn the Game of Facades
Engage Beyond Elections
Culture, Identity & Change: The Evolution of Indian Society
The Triangular Dynamics
Never Trust Without Doubt
If You Can't Change the Country, Leave the Country
Live Fearless with Fear
India: The Investment Magnet
Realignment
Powering a Billion Dreams
Universal Skill
Extension: From Brush to Algorithm
Human Side of Einstein
The Future Knows No Lines
Let's Bring the Generations to the Table
Unified Energy Dynamics
A Beat That Never Dies
Strategic Horizons
Right From Scratch
The Forgotten Innovators
Physics Fixing
The Savior Illusion
The Selfish Side
The Age of the Weakest Sapiens
Neural Wars
The Absorption: There Is No Death
Strange Therapist
Bubbles: Become a Drop
Beyond the Blue
Half-Scholar
The Law of Truth
Failure

Philosophy 2.0: A Structural Manifesto
Chaos: A Gentle Guide to the Unresolvable
Self-Help: New Syllabus
Rippling
Game: How to Play the Share Market
Out of Syllabus
Saffronized: The Code Behind the Crown
The Anger
Moksha: The Liberation
Chavan vs Einstein: The Universe Was Never Curved
Rocket Science: The Ripple Gambling
The Cosmos: The Still Universe
The Taste of Silence
Chavan's Post-Materialist Human Model
10-Second Rule
Chavan vs Hawking
It was Never a Wave Signal
Quantum Courtroom: The Trial of a Century
Chavan–Heisenberg Ripple Uncertainty Principle
The Great Disruption

Watch for more at https://www.sandeepjchavan.com/.

www.ingramcontent.com/pod-product-compliance
Lightning Source LLC
LaVergne TN
LVHW090310210926
848480LV00001B/201
* 9 7 9 8 2 3 1 2 6 5 3 7 4 *